Troubleshooting and Problem-Solving in the IVF Laboratory

Troubleshooting and Problem-Solving in the IVF Laboratory

Kay Elder
Senior research scientist, Bourn Hall Clinic, Bourn, Cambridge, UK

Marc Van den Bergh
Chief Executive Officer, Quartec GmbH, Fislisbach, Switzerland

Bryan Woodward
Freelance consultant clinical embryologist based in Leicestershire, UK

CAMBRIDGE
UNIVERSITY PRESS

CAMBRIDGE
UNIVERSITY PRESS

University Printing House, Cambridge CB2 8BS, United Kingdom

One Liberty Plaza, 20th Floor, New York, NY 10006, USA

477 Williamstown Road, Port Melbourne, VIC 3207, Australia

314-321, 3rd Floor, Plot 3, Splendor Forum, Jasola District Centre, New Delhi - 110025, India

79 Anson Road, #06-04/06, Singapore 079906

Cambridge University Press is part of the University of Cambridge.

It furthers the University's mission by disseminating knowledge in the pursuit of education, learning and research at the highest international levels of excellence.

www.cambridge.org
Information on this title: www.cambridge.org/9781107673175

First published 2015

A catalogue record for this publication is available from the British Library

ISBN 978-1-107-67317-5 Paperback

This book is dedicated to the memory of Bob Edwards (1925–2013) and Jean Purdy (1945–1985), as a tribute to their 10 years of heroic and persistent IVF troubleshooting in Oldham, 1968–1978. Their wisdom, courage, and indefatigable enthusiasm and stamina in facing and overcoming multiple problems, setbacks, and challenges was finally rewarded with the birth of the first two Test Tube Babies, both conceived towards the end of their troubleshooting journey in Oldham: Louise Brown (single embryo transfer November 12th 1977, born July 25th 1978) and Alastair MacDonald (single embryo transfer May 23rd 1978, born January 14th 1979).

Contents

Contributors

Kimball O. Pomeroy, PhD, HCLD
South Florida Institute for Reproductive Medicine, Miami, FL, USA

Michael Reed, PhD, ELD HCLD (ABB)
Center for Reproductive Medicine of New Mexico, Albuquerque, NM, USA

Preface

In September 1998, approximately 200 IVF scientists worldwide participated in a 48-hour International Internet Conference hosted by Thomas Elliott and Kay Elder on behalf of Alpha, Scientists in Reproductive Medicine: "Problem Solving and Troubleshooting in IVF." The edited Proceedings of this Conference were published by Phil Matson and Ladybrook Publishing in 1998, as one of a series of World Wide Conferences on Reproductive Biology; copies of this book ran out within a month of its printing, and it remained out of print thereafter. Demand for the book continued, and electronic pdf copies were sent on request over the subsequent decade. Fourteen years later, in 2012, Dr. Rajvi Mehta requested permission to re-print the book in order to help local embryologists in India, with the support of Trivector Scientific Pvt Ltd and Trivector Embryology Support Academy. This confirmation of continuing demand for guidance in troubleshooting IVF systems prompted us to consider writing an update to the original Conference Proceedings book.

During 2013, Quartec GmbH launched a series of Troubleshooting workshops held in a number of European countries, along with a series of scheduled Webinars (www.quartec.co); the popularity of both workshops and webinars further reinforced the suggestion that an updated book would be a useful contribution to the field of IVF Clinical Embryology. A third co-author (BW), agreed to join us, with the valuable contribution of practical experience, case reports, and tips gained from troubleshooting in IVF laboratories all over the world.

The chapters include commonly encountered problems, supported by case reports, and contain the essential theoretical background necessary to understand the science and implement solutions. Each chapter contains specific practical tips, as well as troubleshooting flowcharts and process maps that can be followed in tackling different types of problem. The scope of the book is intentionally restricted to routine clinical IVF/ICSI practice only: more complex and novel technologies can only succeed if they adhere to the same established fundamental basic principles that contribute to success in conventional culture of gametes and embryos.

We hope this book will provide a useful guide in helping IVF clinics to maintain the highest, consistently reliable success rates possible, in order to achieve the dream of every infertile couple: the birth of a healthy live baby.

Acknowledgments

We are grateful to all of the participants of the 1998 International Internet Conference on Troubleshooting and Problem Solving in IVF for their many valuable contributions at that time. Topics and questions posted on Embryomail (www.embryomail.net), and on IVF.net (www.ivf.net) provided a further valuable resource and source of inspiration, and we are immensely grateful to all of the scientists who continue to post questions, suggestions, and advice with such enthusiasm and dedication. Thanks also to Nadir Çiray and Yves Ménézo for helpful comments and suggestions.

Kay gratefully acknowledges the continuing support of Mike Macnamee and all of the team at Bourn Hall Clinic. . .. Thank you! Bryan would like to acknowledge all of the staff in laboratories where he has been called in to provide locum embryology, consultancy, and troubleshooting. He is also grateful to all of the labs that permitted photographs to be taken.

Chapter 1

Introduction

Fighting against Murphy's Law in ART

"If something can go wrong, it will…": Murphy's Law has belonged to popular wisdom for many years. The modern version of the law has its roots in 1949 US Air Force studies on the effects of rapid deceleration on pilots: volunteers were strapped on to a rocket-propelled sled, and their condition monitored via electrodes fitted to a harness as the sled was brought to an abrupt halt. The monitoring harness was designed by Captain Edward A. Murphy, and technicians were puzzled by the fact that it failed to record any data. Murphy discovered that all of the electrodes had been wired incorrectly, prompting him to declare:

> *"If there are two or more ways of doing something, and one of them can lead to catastrophe, then someone will do it."*

Project engineers presented Murphy's observation as an excellent working assumption in safety-critical engineering, but to Murphy's chagrin, his principle was transformed into an apparently flippant statement about the cussedness of everyday events. Ironically, by losing control over his original meaning, Murphy thus became the first victim of his eponymous law. Murphy was perhaps unaware that the essential idea behind his law had been observed by the Scottish poet Robert Burns in 1786:

> *"The best laid schemes o' Mice an' Men Gang aft agley (often go awry)."*

Another famous example of Murphy's law was described in 1884 by the Victorian satirist James Payne:

> *"I never had a piece of toast*
> *Particularly long and wide*
> *But fell upon the sanded floor*
> *And always on the buttered side."*

Many scientists will consider Murphy's law as part of a vast body of folklore, considered both humorous – and a source of truth about human nature. Robert Matthews, in a scientific study of the Tumbling Toast phenomenon (1996) stated that the awful truth is that the universe is against us. He used a wide range of mathematics and science, from probability theory to rigid-body dynamics and found that Murphy's law finds its explanation in cosmic constants, combinatoric analysis, etc.

Perhaps Murphy's true legacy is the concept that *apparently trivial phenomena often have explanations that are far from trivial.* This brings us back to a basic premise: *Life is difficult!* and to the first of Buddha's Four Noble Truths, which states: *"Life is suffering."* The embryologist may add to this: *"ART is more suffering."* Applying M. Scott Peck's vision (1978) we may state:

> "Once we truly know that ART is difficult – once we truly understand and accept it – then ART is no longer difficult, because once the principle is accepted, the fact that it is difficult no longer matters."

Peck explains that problems cause uncomfortable feelings such as frustration, grief, guilt, anger, fear, and anxiety. Nevertheless, problems are also the

> "cutting edge that distinguishes between success and failure. Problems call forth our courage and our wisdom: indeed, they create our courage and our wisdom."

The field of human ART evolved over a "gestation period" of nearly 10 years, in Oldham, a small town in the north of England, near Manchester. After starting their laboratory work in a laundry room adjacent to the operating theatre at Oldham and District General Hospital, a small room in nearby Dr. Kershaw's Cottage Hospital (Fig. 1.1) was converted into a makeshift laboratory where Bob Edwards and his assistant Jean Purdy developed the methods that were to lead eventually to successful IVF. The arduous journey undertaken by Robert Edwards, Patrick Steptoe, and Jean Purdy that led to the birth of Louise Brown in 1978 (Fig. 1.2) was beset with problems from every possible perspective: clinical, laboratory, logistic – as well as controversy and criticism from numerous aspects of the outside world. There is no doubt that their eventual success certainly required exceptional – and extraordinary – courage and wisdom (Edwards & Steptoe, 1975; Elder & Johnson, 2015a, b, c; Johnson & Elder, 2015a, b, c), and a study of their journey through numerous challenges and problems represents a classical approach to troubleshooting. Their empirical approach to treatment variation was based on several previous decades of background research and experience, so that they were able to derive a rational approach to the many problems encountered. Without recourse to any of the sophisticated technology that subsequently became available, this crucial background of accumulated wisdom allowed each problem encountered to be addressed scientifically, using methods that were solidly based on informed reading of the contemporary literature available at the time.

Despite nearly four decades of progress in reproductive medicine and technology, problems continue to be inevitable; our tools have increased in complexity, and complex systems bring problems that can only be tackled with an in-depth knowledge of the associated scientific principles. Computer-controlled data acquisition and continuous monitoring systems can provide records from multiple instrument inputs, but the data must be checked and analyzed by well-trained, educated, and experienced scientists who are able to appreciate that *apparently trivial phenomena often have explanations that are far from trivial.* This is especially true in ART, where small details that might be perceived as insignificant can collectively contribute to successful healthy embryo development. Problem-solving to dissect and discover the root and cause, whether or not it may be "far from trivial," requires an understanding of the many background layers that make up the sophisticated physico-chemical and physiological foundation of IVF systems.

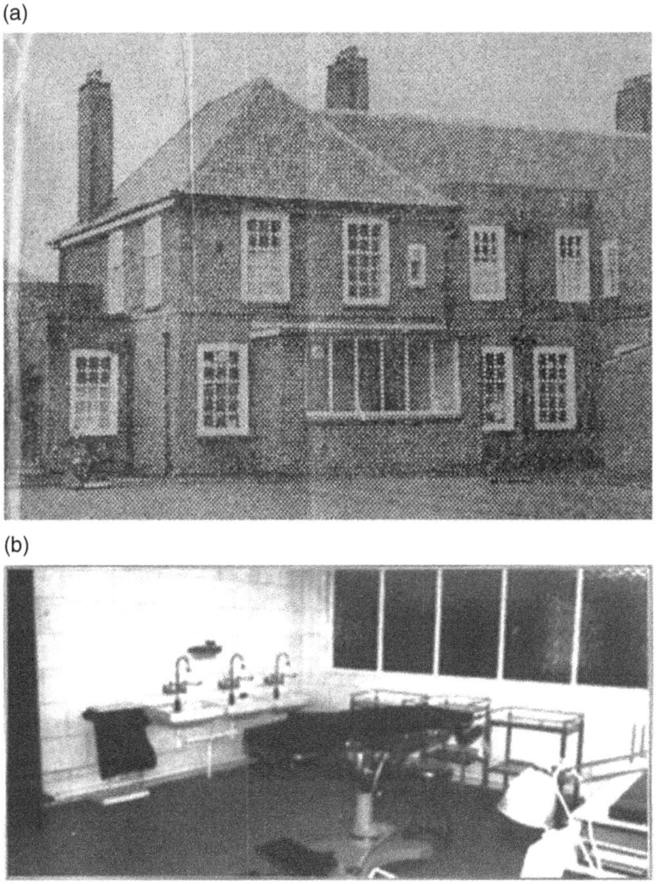

Fig. 1.1 Dr. Kershaw's Cottage Hospital, near Oldham, Manchester.

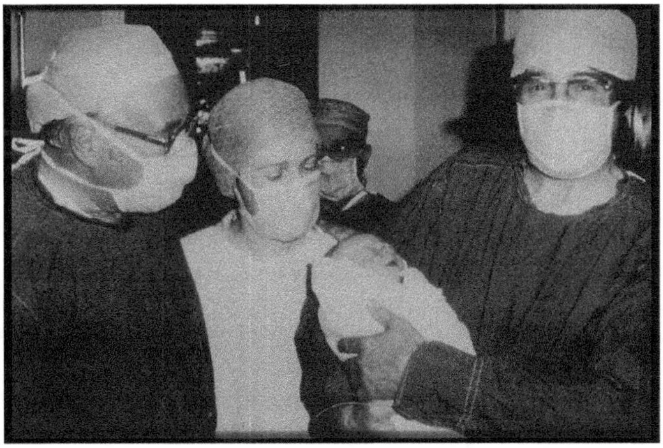

Fig.1.2 The birth of Louise Brown at Oldham and District General Hospital, July 25th 1978; Patrick Steptoe, Jean Purdy (center) and Robert Edwards holding the baby.

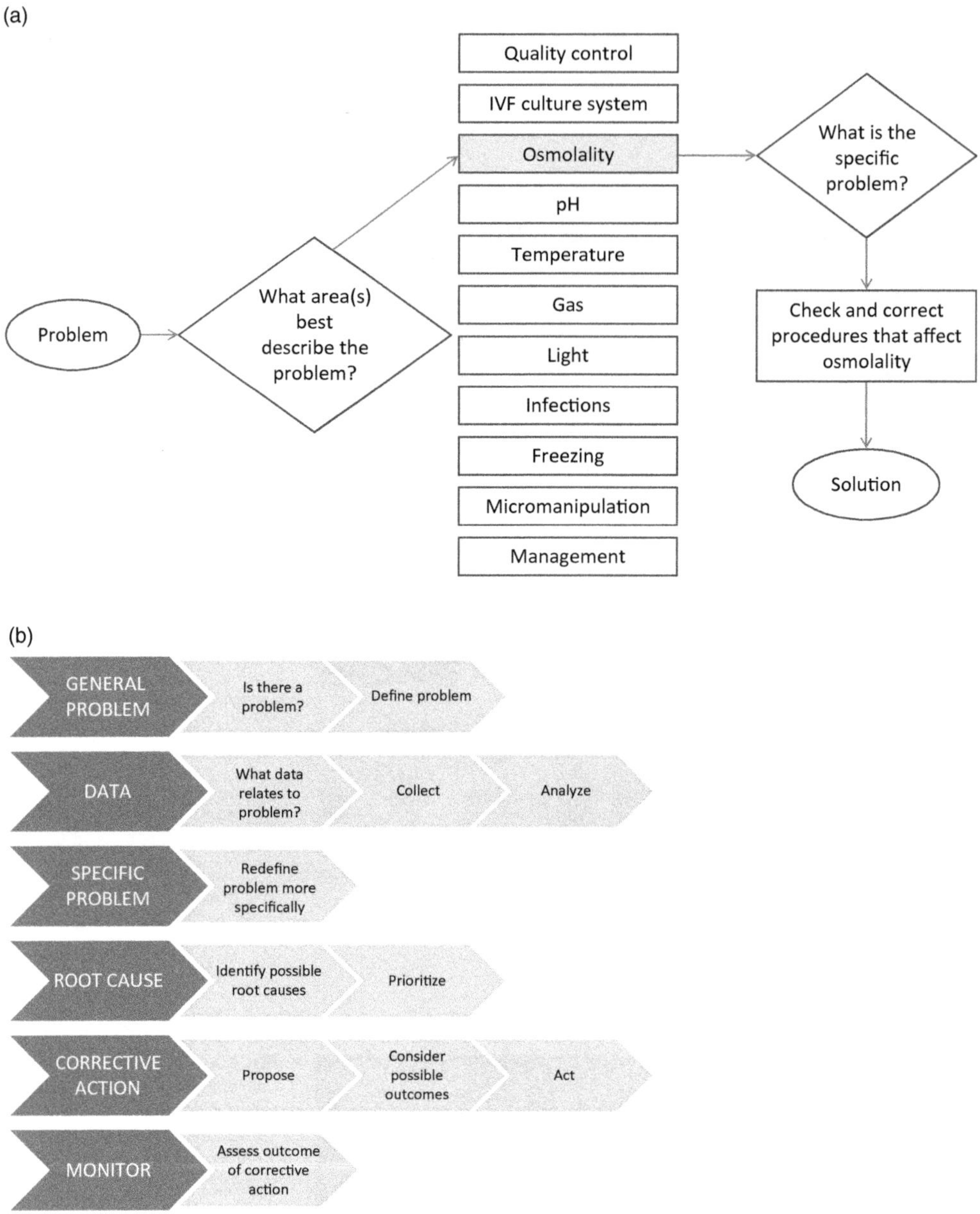

Fig. 1.3 (a) General troubleshooting process map. (b) Troubleshooting guide.

Hoping that ART will be a "bed of roses" is setting ourselves up for failure, as every rose has its thorn. A carefully controlled scientific approach to every detail of the complex procedures involved in ART can help to reduce the impact of Murphy's law on the final outcome of the program: the live birth of healthy children.

"Troubleshooting" is defined as a form of problem-solving, applied to repair failed products or processes. It is a logical, systematic search for the source of a problem in order

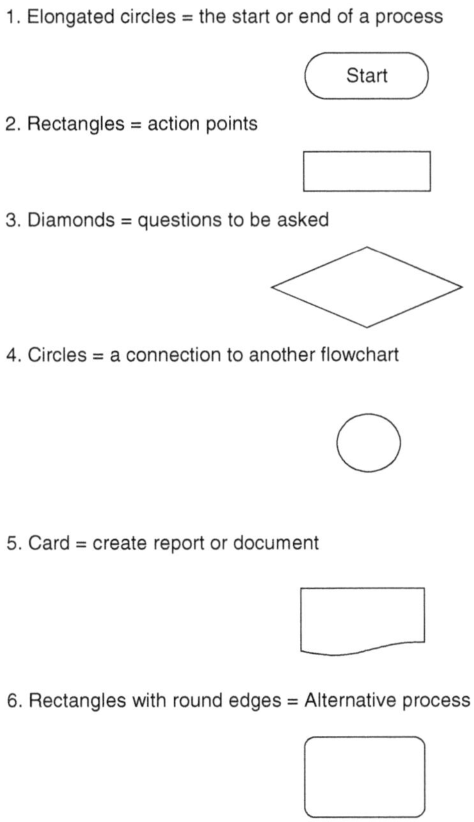

Fig. 1.4 Flowchart symbols: each symbol can be accompanied by a very brief summary that indicates the start or end of the process, the action to be taken, and the decision to be made.

to find a solution that will make the product or process successfully functional again. The ART laboratory represents the synthesis of multiple complex systems, and the symptoms of a problem can have many possible causes. Attention must be paid to any element in the culture system that might cause physiological stress to embryos, reducing their potential viability. Effective maintenance of any ART laboratory requires an ongoing process that will support troubleshooting.

It is often tempting to blame external supplies for any downturn in results, when the real problem may be something quite different. In order to make a valid assessment of a potential problem, it is essential to keep consistent and reliable records at all times, so that parameters can be compared during different time periods. Rates of fertilization, cleavage, blastocyst development, implantation, biochemical and clinical pregnancy should form a baseline for comparison. Any changes in patient demographics, including age, fertility diagnoses, stimulation protocols, drug batches and doses, response to stimulation, sperm characteristics, and oocyte quality for embryos cultured can have downstream effects on results, as well as any changes in culture conditions: incubator(s), measured temperature and pH, gases, dishes, and the culture media (basic constituents and supplemented protein).

IVF systems as a whole are made up of several interrelated and overlapping complex systems, each with an impact on other factors within the system. Gamete quality is affected by the patient's environment and lifestyle, as well as by clinical features during ovarian

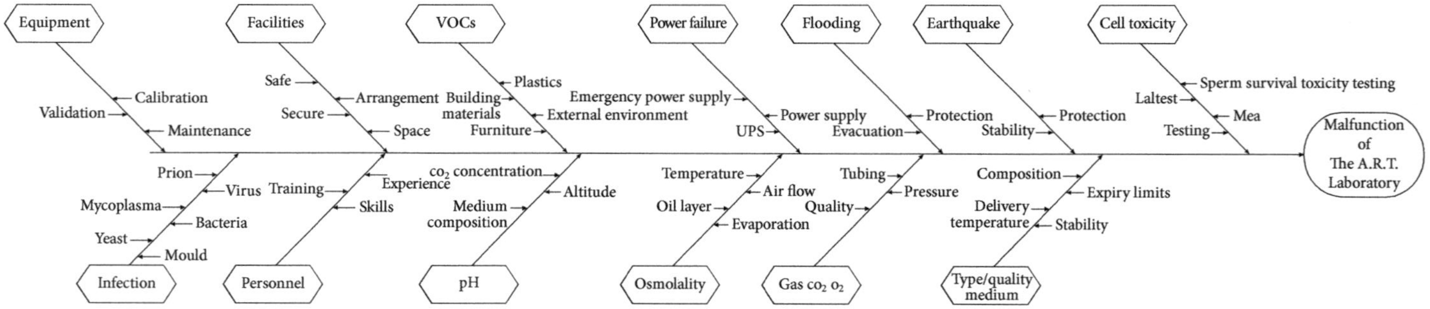

Fig. 1.5 General troubleshooting flowchart.

stimulation, monitoring, and oocyte retrieval. Within this perspective, providing the gametes and embryos with optimal and stable conditions within the laboratory culture system represents further interrelated and overlapping parameters that can easily go wrong. When problems arise, it can be very difficult to know where to begin. The process of troubleshooting can be rationalized/simplified by a summary of steps, using flowcharts to provide guidance and direction to potential problems. Concentrating on basic individual steps allows more precise focus, lessening the overwhelming impact of the general problem within the big picture:

1. Define the problem
2. Produce a step-by-step process for analysis
3. Re-define the problem, find the root cause, and suggest solutions to the problem (Fig. 1.3).

Flowcharts that are easy to follow can be built using symbols that are linked by arrows to show the direction of flow (Fig. 1.4).

Examples of suggested troubleshooting process maps and flowcharts are provided at the end of each chapter in this book.

Clinical embryologists who handle human embryos do not "create" life, but instead have the responsibility of "nurturing" life by protecting the vulnerable preimplantation embryo from physiological and biochemical stress factors associated with *in vitro* culture. All technical procedures must be handled in a manner that respects and supports the physiology, biochemistry, chemistry, and physics that allow components of the culture system to interact harmoniously. We hope that the following chapters will provide guidance towards maintaining a homeostatic culture environment that will nurture the potential viability of each precious human embryo entrusted to our care (Fig. 1.5).

Further reading

Go KJ, Patel JC, Dietz R (2012). Troubleshooting in the clinical embryology laboratory: the art of problem-solving in ART. In: *Practical Manual* of In Vitro *Fertilization, Advanced Methods and Novel Devices*, P Nagy, AC Varghese, & A Agarwal (eds.) Springer NY, p.631–7.

Mortimer D, Mortimer ST (2007). *Quality and Risk Management in the IVF Laboratory.* Cambridge University Press.

References

Edwards RG, Steptoe PC (1975). Induction of follicular growth, ovulation and luteinization in the human ovary. *J Reprod Fertil (Suppl)* **22**, 121–63.

Elder K, Johnson MH (2015). The Oldham Notebooks: an analysis of the development of IVF 1969–1978. *Reprod Biomed and Society, in press.*

Johnson MH, Elder K (2015). The Oldham Notebooks: an analysis of the development of IVF 1969–1978. *Reprod Biomed and Society, in press.*

Mathews RAJ (1995). Tumbling toast, Murphy's Law and the fundamental constants. *Eur J Phys* **16**, 172–8.

Peck MS (1978). *The Road Less Travelled*, Simon & Schuster, NY.

Chapter 2

Quality control, certification, and accreditation

Introduction: guidelines and good practice

Careful Quality Control (QC) in the IVF laboratory has always been recognized as essential for success. Controversy surrounding the birth of Louise Brown in July 1978, together with lack of funding, brought the work of Steptoe and Edwards to an abrupt halt in 1978 (Johnson *et al.*, 2010), to be resumed only after Bourn Hall opened its doors to patients in September 1980. The patients were prepared for natural cycle IVF, using the same methods developed in Oldham (Elder & Johnson 2015a, b, c), and the first laparoscopic oocyte retrieval was carried out in October 1980. Over the following 2-week period, two out of nine embryo transfers led to pregnancies and subsequent healthy live births: a remarkable success, which could only have been achieved by meticulous attention to detail: "Quality Control" – at every step. The first international IVF meeting took place at Bourn Hall, Cambridge, UK, in 1981 (Fig. 2.1), with proceedings of the meeting published in 1982 (Edwards & Purdy, 1982). A chapter authored by Jean Purdy describes some of the difficulties that the Oldham team had encountered in various stages of the procedure between 1969–1978, and clearly states (Purdy, 1982; p.135):

"It is essential to maintain a careful and rigorous control over the methods used throughout each stage, the purchasing and testing of equipment, methods used in making and testing media, preparing spermatozoa, and achieving fertilization and embryo culture in vitro."

Successful ART continues to be crucially dependent on carefully controlled conditions in every aspect of the IVF laboratory routine, and a comprehensive total Quality Management System (QMS) is mandatory. An accredited/certified laboratory is obliged to provide evidence that all aspects of facilities, treatment, and personnel meet the standards required by specific regulatory bodies. Organizations such as the Human Fertilization and Embryology Authority (HFEA) in the UK, the College of American Pathologists (CAP) in the USA, and Joint Commission International (JCI) require documented evidence of a QMS. In Europe, all EU States have been asked to incorporate the European Union Tissues and Cells Directive 2004/23/EC into their country's regulation for IVF. This Directive, issued in 2004, sets standards of quality and safety for the donation, procurement, testing, processing, preservation, storage, and distribution of human tissues and cells. The International Organization for Standardization (ISO: www.iso.org) provides sets of standards, which include:

- ISO 9000:2000. Quality Management Systems: Fundamentals and vocabulary
- ISO 9001:2008. Quality Management Systems: Requirements
- ISO 15189:2012. Medical laboratories: Requirements for quality and competence.

Fig. 2.1 Group photograph of the First IVF meeting, held at Bourn Hall Sept 5th – 8th 1981; front row: Bob Edwards, Jean Purdy, Patrick Steptoe, John Webster, Simon Fishel.

The ISO also produces standards in collaboration with the International Electrotechnical Commission (IEC), which includes ISO/IEC 17025:2005: an innovative standard for laboratory quality systems. ISO standards are endorsed by the European Committee for Standardization (CEN: www.cen.eu) which describes Quality Assurance systems.

Elements of a quality management system

Accurate and comprehensive records are the basic foundation for troubleshooting, the essential first step that allows any potential hazard that may have been introduced to be detected. Components of a good QMS will cover not only a basic audit of implantation and pregnancy rates, but also reflect the accuracy and reproducibility of results, as well as confirming appropriate standards for the largest possible range of items related to laboratory practice. Systematic and periodical surveys assess:

- Optimum function of all equipment and instruments in use
- Quality of all consumables and media in use
- Critical periodic analysis of individual performance within the team.

Essential elements of a QMS include:

1. Process flowchart
2. Standard Operating Procedures (SOPs)

3. Job descriptions for all members of staff and an organizational chart
4. Forms: for information, records of procedures, instructions, etc.
5. Consent forms
6. Key Performance Indicators (KPIs).

Standard Operating Procedures (SOPs), written in a consistent format, describe protocols for all procedures carried out in the laboratory; other areas, e.g. nursing, clinicians, administration, etc. will also have their own SOPs. SOPs undergo a system of scheduled review, ensuring that personnel are familiar with the SOPs for all routinely performed procedures, as well as with updates following each review. The accepted SOP format includes:

- The goal of the procedure
- Personnel allowed to use and apply the procedure
- Products and equipment used in the procedure
- How to perform the procedure
- Validation for the procedure
- Any other SOPs and forms related to the described procedure.

Equipment specifications

A user's manual is essential for every piece of equipment, together with traceable validation, e.g:

- Installation Qualification (IQ) – documented proof that the equipment has been delivered and installed in accordance with the requirements and statutory safety regulations stipulated in the design qualification. IQ documentation includes a test plan and a report.
- Operation Qualification (OQ) – documented proof of a test process that evaluates whether the equipment functions correctly. All items specified in the test plan are processed and documented in writing, to ensure that the system functions in accordance with specifications. OQ documentation includes a test plan and a report.
- Performance Qualification (PQ) – documented proof that verifies reproducible equipment function across the entire specified working range and limits.

All maintenance records are dated and kept in a QC register, including annual service (or less or more frequently, as required). Documented records of the working parameters are kept, e.g. logs of incubator temperature and gas composition, logs that document the first used of each batch of new items and supplies, as well as usage logs for new or repaired equipment.

Monitoring laboratory equipment

The chance of equipment failure or suboptimal operation can be minimized by establishing service contracts with reliable companies. ISO norms require that companies are able to provide calibration certificates, including a description of the method used for calibration and the tools used in servicing and calibrating the equipment. Metrological traceability of the calibration standard for each item of equipment is required, and any correction factors need to be accounted for in future use. The calibration status and recommended re-calibration date should be clearly recorded. Equipment manuals for all equipment should

Box 2.1 Case report

During preparation of a sperm sample for IVF, the centrifuge lid became jammed at the end of the centrifugation, with the display indicating that it was still running. The sample was due for insemination, the embryologist did not know what to do, began to panic, and called for help in troubleshooting. The first step was to locate the equipment manual, which contained a troubleshooting section. This showed the location of an emergency "unlock" switch, at the base of the lid. Had the embryologist been familiar with the manual, this problem could have been easily resolved. The root cause of the lid "sticking" then had to be investigated, which required a visit from the manufacturer.

(Note: some centrifuges require a special key; it is advisable to keep this fixed to the side of the centrifuge.)

be readily accessible; individual SOPs written specifically for each item of equipment may also be useful. A manual may have a troubleshooting section, which can be essential in case of a problem (Box 2.1).

Contracts for service and documented calibration are required for:

- Incubators
- Incubator alarms
- Flow cabinets
- Microscopes
- Micromanipulator stations
- Heated surfaces/microscope stages
- Centrifuges
- Refrigerators and freezers
- Embryo/oocyte freezing machines
- Osmometers
- Liquid nitrogen storage dewars and vapor freezers
- Low-level nitrogen and oxygen alarms in the cryostore
- Air filtration equipment
- Electronic witnessing system
- Time-lapse embryo monitoring systems
- Automatic piston micropipettors (calibration and servicing).

All vital equipment should have alarm systems, with provision for emergency call-out. Online alarm systems are effective only if they result in immediate intervention (within minutes) or via direct online access. Key items of equipment in daily use also require systems of continuous monitoring to ensure optimal performance, with carefully chosen measuring tools that are appropriate for the task and correctly calibrated: an apparent malfunction can be due either to the equipment, or to the measuring tool.

IVF laboratories can now be equipped with a master control system that continually monitors the performance of all critical equipment and provides alarm notifications if there is a failure. Figure 2.2a shows a plan of an IVF laboratory with locations of critical equipment, and a home screen showing real-time values and status for each piece of equipment monitored. Fig. 2.2b shows an alarm notification where a refrigerator

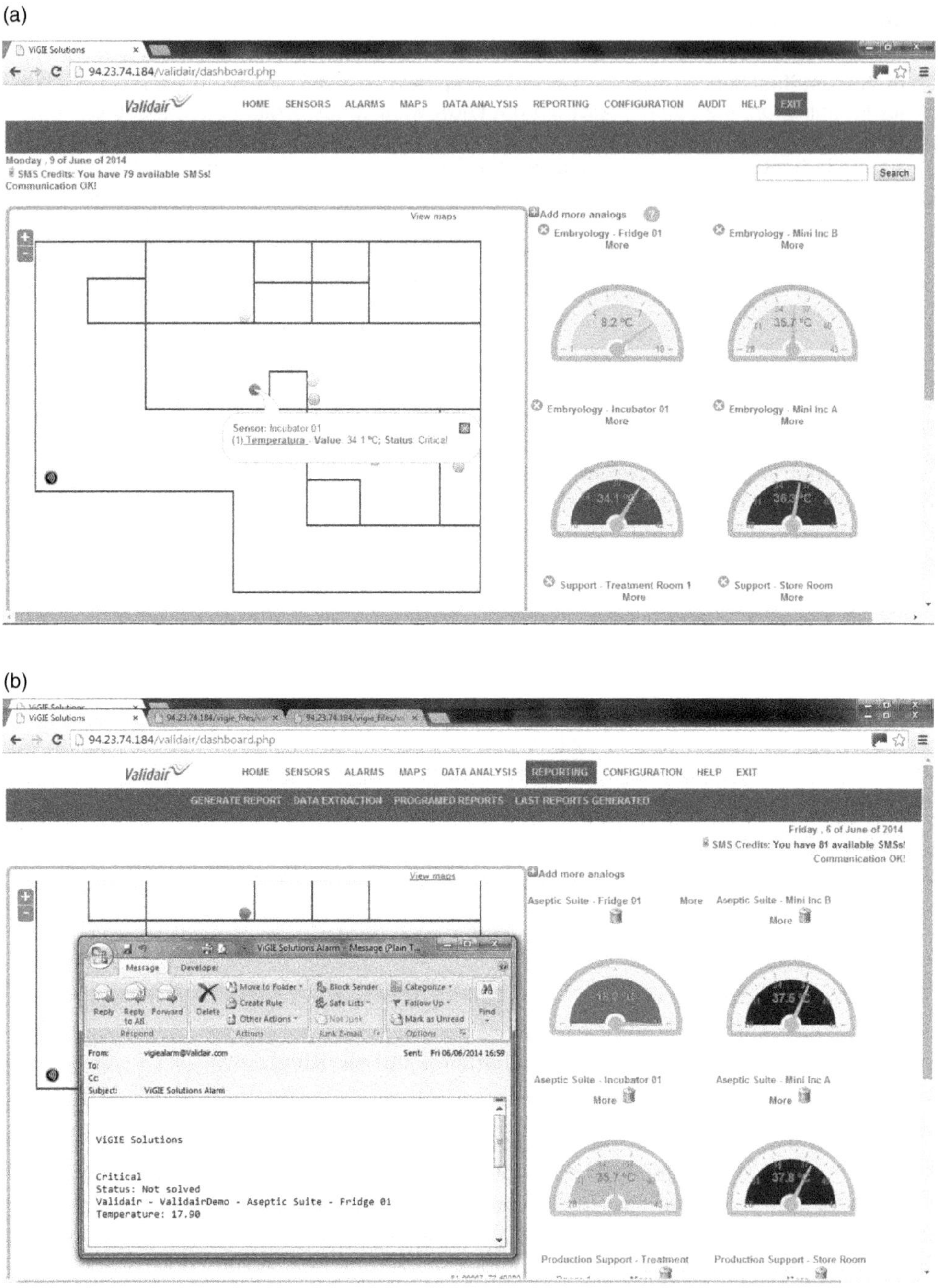

Fig. 2.2 QC monitoring in an IVF laboratory. *(Images from their demanstration system kindly provided by Mr Dave Sears, Validair, Kent, UK.)* (a) Plan of an IVF lab with locations of the critical equipment, and a home screen showing real-time values and status for each piece of equipment monitored. (b) Alarm notification where a refrigerator temperature range has reached a critically high level. An SMS autodialler alerts the embryologist "out of hours" by telephone and an email provides specific details of the cause of the alarm.

(c)

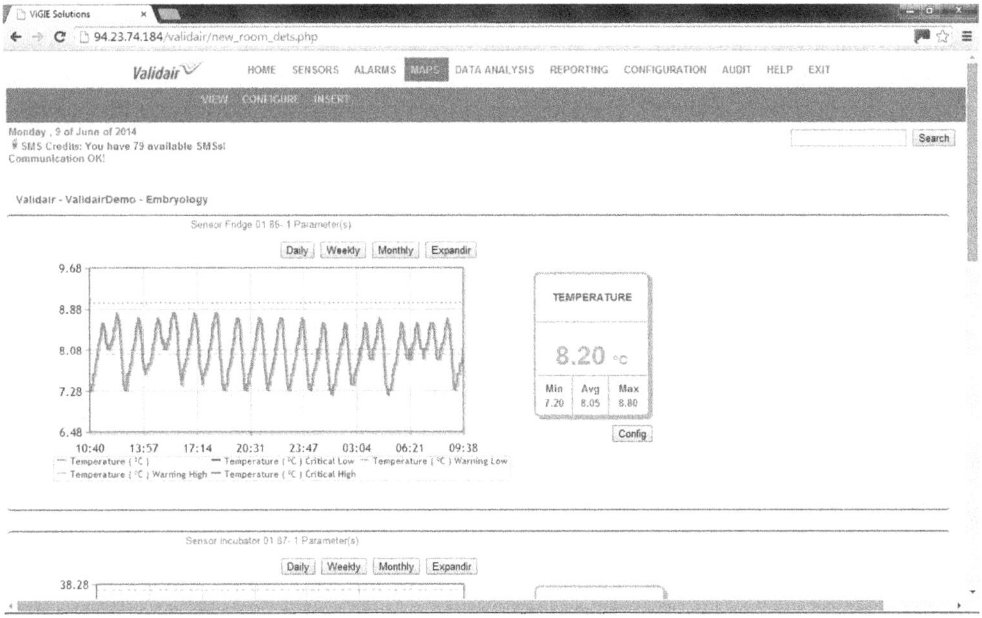

(d)

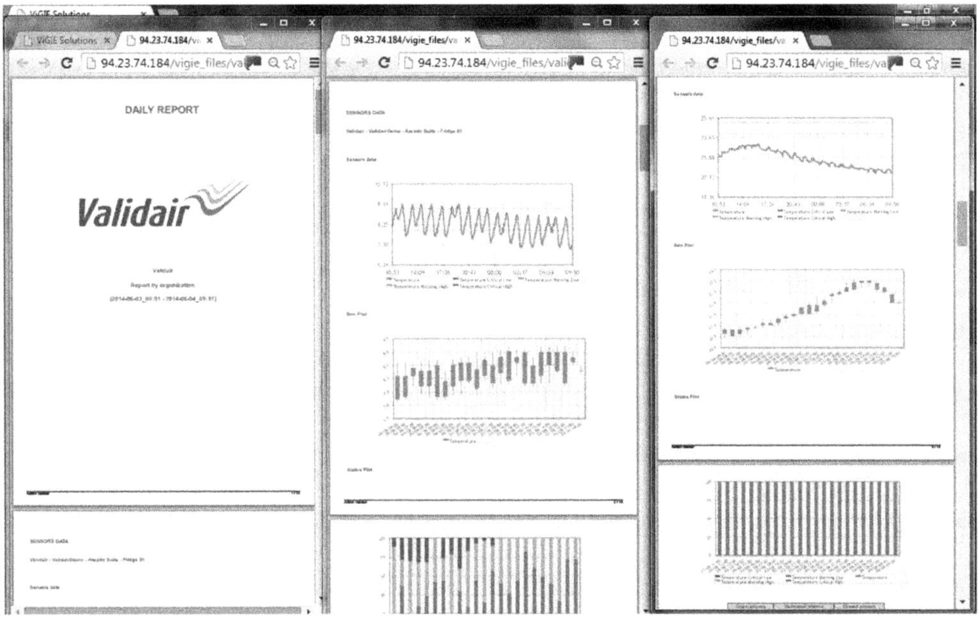

Fig. 2.2 *(cont.)* (c) Historical data represented in graphical format, with current values and status. (d) Example of an automated daily report showing some of the graphical formats available.

temperature range has reached a critically high level; an SMS autodialler alerts the embryologist "out of hours" by telephone, and an email provides specific details of the cause of the alarm. Fig. 2.2c presents historical data in graphical format, with current values and status. Fig. 2.2d is an example of an automated daily report showing some of the graphical formats available. (Images from their demonstration system kindly provided by Mr. Dave Sears, Validair, Kent, UK.)

A consistent, reproducible culture environment depends on clearly defined and tight tolerance levels, established individually for each item or system that is monitored in the laboratory. Remove defective equipment from service immediately. If it is too large to remove from the laboratory, label it clearly as "out of action – do not use," and inform all laboratory staff that it is defective. Document details of the make, model, serial number, date of acquisition, and date of last service in an equipment service record book. Log the date of the fault and contact the service company or manufacturer to instigate a repair; fully decontaminate all equipment before either an engineer visit, or sending it off-site for repair. The age and lifespan of the equipment also needs to be considered, in case replacement is preferable. Following repair (or replacement), sterilize the equipment fully prior to (re)introduction to the lab, and verify performance (with full documentation for traceability) prior to use. If new, the manufacturer should provide the IQ, OP, and PQ. Document all details in the laboratory equipment records.

Consumable products and supplies

Use disposable items that are guaranteed as non-toxic and tissue culture grade: needles for oocyte aspiration and the collecting system, manipulation pipettes and pipette tips, culture vessels (dishes, tubes, etc.), and catheters for insemination or embryo transfer. Disposable pipettes used for gamete and embryo manipulations can be purchased pre-sterilized and packaged. Since manipulation pipettes with QA certificates are available over a range of lumen sizes, these are recommended over pipettes that are "hand-pulled" over a flame (this process invalidates the QA of the original sterile Pasteur pipettes). Important considerations in the selection of supplies include:

- Suitability for intended purpose
- Suitable storage facilities (i.e. spacious storerooms away from excessive heat/direct sunlight); items that have been sterilized can continue to release volatile organic compounds for some time, and a storage area that is independent from the laboratory working areas is a useful precaution.
- Compliance of suppliers to a contract with specified terms and conditions, i.e. Third Party Agreements
- Delivery of perishable items such as media under controlled conditions (e.g. maintaining a specific temperature during transit)
- Batch numbers to be recorded for QA purposes; check carefully to confirm that the batch number on the packaging matches the batch number on the item (note that some countries have a unique batch number for boxes that differs from the contents. This is acceptable provided that all items are traceable).
- Disposable plastics are CE-marked (for Europe), FDA approved (for the US), MEA tested where possible and certified as endotoxin-free with validation evidence. DNA-RNAse free pipette tips are recommended (NB: CE-marking is *not* a guarantee of

quality, but only of conformity in manufacture; it indicates that production was carried out according to established rules in Europe, and does *not* guarantee that the product has been tested in cell culture and is non-toxic).

- All media should be supplied with Certificates of Analysis; keeping these Certificates allows traceability when necessary.

Bioassays

Manufacturers of commercial media and companies who distribute disposable supplies include a Certificate of Analysis with each shipment, indicating the type and results of bioassays used in their QC testing. Companies that supply commercial IVF culture media are obliged to carry out a number of QC procedures that include testing raw and contact materials before release for media manufacture, in-process testing, and testing of finished products prior to release. Certificates supplied with each batch normally confirm that all physico-chemical parameters (pH and osmolality) are correct, that the batch has been tested for sterility, and also include results of bioassay procedures such as sperm viability test or MEA. However, the ability to carry out independent in-house tests on goods that are commercially supplied is essential in troubleshooting situations. Testing new batches of media, oil, material, or supplies used in the culture system before use is recommended. The physico-chemical limits of culture media testing are crucial:

- Osmolality within the limits of 275–305 mOsm/kg, with a total variation of no more than 30 mOsm/kg (see Chapter 4)
- pH within the limits of 7.2–7.5, with a maximal variation of 0.4 units of pH (see Chapter 5).

Larger variations in either parameter may indicate poor technique/technology and/or inadequate controls during manufacture, with potential impact on clinical results.

Tissue culture plasticware may be subject to variation in quality: well-to-well variations have been observed even within a single four-well plate, and rinsing plates with media before use may be a useful precaution. Carefully inspect all plasticware for damage prior to use, as fragments of plastic or cracks can expose embryos to toxins – a dish surface may be coated with an "embryo-safe" layer, but the coating will be compromised if the layer is damaged. Manufacturers of tissue culture plasticware may change the chemical formulation of their products without notification, and any changes in manufacture of syringes, filters, and culture dishes may sometimes be embryotoxic. It is also important to store plasticware at an ambient room temperature, away from direct sunlight, as this can affect the properties of the plastic and the packaging. Some plasticware carries a scent or odor, caused by emission of volatile organic compounds (VOCs) (Hall *et al.*, 1998; see Chapter 8). Chemicals such as styrene, ethylbenzene, and benzaldehyde have been recorded for plastic ware at significantly high levels (Gilligan *et al.*, 2008; Holyoak *et al.*, 1996). It is important to allow plasticware to off-gas prior to use. Although the manufacturer is responsible for ensuring that products are off-gassed in a sterile environment prior to delivery, any detectable scent or odor after the packaging is opened in the laboratory should be noted, and dealt with.

Guidelines issued by regulatory bodies suggest that media and supplies used in the IVF laboratory are QC tested with an appropriate bioassay system, but the choice of an appropriate system for human gametes and embryos continues to be a matter of debate.

Box 2.2 Mouse embryo assays

The validity of a MEA has been questioned as a reliable assay for extrapolation to clinical IVF: it assumes that the requirements of human and mouse embryos are the same, and this is a false assumption. The mouse embryo cannot regulate its endogenous metabolic pool before the late two-cell stage; this is not the case for human or bovine embryos. Mouse embryos will develop from the two-cell stage onwards in a wide variety of cell culture media (George *et al.*, 1989). In a 4-year retrospective analysis of culture medium QC using mouse IVF and mouse zygote bioassays, Van den Bergh *et al.* (1996) concluded that media showing high scores with the MEA did demonstrate higher fertilization rates and better embryo quality when used for human IVF. However, the success of any MEA is dependent upon the mouse strain used, the stage at which the assay begins (one-cell or two-cell) and upon the type of mouse embryo culture used, with rigorously controlled and reproducible procedures. Batches of oil that are toxic to human embryos have often been known to "pass" an MEA. In addition, culture of mouse embryos is not readily accessible to many routine human IVF laboratories.

A range of bioassays to detect toxicity and suboptimal culture conditions have been used in human IVF (Rinehart *et al.*,1988), including:

1. Human or hamster sperm survival (Critchlow *et al.*, 1989; Nijs *et al.*, 2008; Hossain *et al.*, 2010)
2. Continued culture of single or multipronucleate zygotes
3. Mouse hybridoma cell line (Bertheussen *et al.*, 1989)
4. Limulus Amoebocyte Lysate (LAL) test for endotoxins (Blechova & Pivodova, 2001)
5. Mouse embryo assays (MEA): culture of mouse zygotes or embryos to the blastocyst stage (Box 2.2).

Key points in the use of bioassays

Although none of the systems currently available can guarantee that subtle levels of toxicity will be detected, they can be helpful in providing baseline data for comparative purposes, and for identifying specific problems. Any bioassay performed routinely and frequently with baseline data for comparing deviations from the norm will be helpful in minimizing the chance of randomly introducing contaminants into the system, and is a useful investment of time and resources in an IVF laboratory. Ultimately, a routine proactive bioassay may be preferable to reactive troubleshooting on the basis of clinical results.

Key points for consideration include:

- Choice of bioassay
- Selection of materials to test
- Frequency of testing
- Establish a set of standards, routine and schedule
- Establish acceptable performance range
- Document all results
- Review results regularly
- Do not use anything that fails the bioassay
- Write an SOP for any corrective action required.

Useful routine bioassay procedures

Sperm viability test

Select a normal sample of washed prepared sperm and assess for concentration, motility, and the type of progression (rapid and slow). Divide the selected sample into four aliquots: add test material to two aliquots, and equivalent control material (in current use) to two aliquots. Incubate one control and one test sample at 37 °C, and one of each at room temperature. Assess each sample for motility and progression after 4, 8, 12, 24, and 48 hours (a computer-aided system can be used if available). Test and control samples should show equivalent survival. If there is any doubt, repeat the test. The Sperm Viability Index (SVI) represents the percentage of progressive motile sperm in the test sample divided by the percentage of progressively motile sperm in the control sample at the specified times (Nijs *et al.*, 2009) (Fig. 2.3).

Single or multipronucleate embryo culture

Oocytes that show abnormal fertilization on Day 1 after insemination can be used for testing new batches of material. Observe, score, and assess each embryo daily until Day 6 after insemination.

Culture of surplus embryos

Surplus embryos after embryo transfer that are not suitable for freezing have also been used for testing new culture products, but this practice is not allowed under an HFEA treatment licence in the UK; the HFEA Code of Practice, Ch(94)09 states: "*An embryo may not be kept under a treatment licence if it is no longer to be used in the treatment of a woman, or if any information that may be obtained from keeping it will not be of specific use in the treatment or diagnosis of the individuals whose gametes were used to produce it. If there is some specific value for a couple in continuing to culture spare embryos from a particular treatment cycle, this should be done with the knowledge and agreement of the couple concerned.*"

The HFEA Act as amended in 2008 allows surplus embryos to be used for training purposes, provided that written patient consent is obtained.

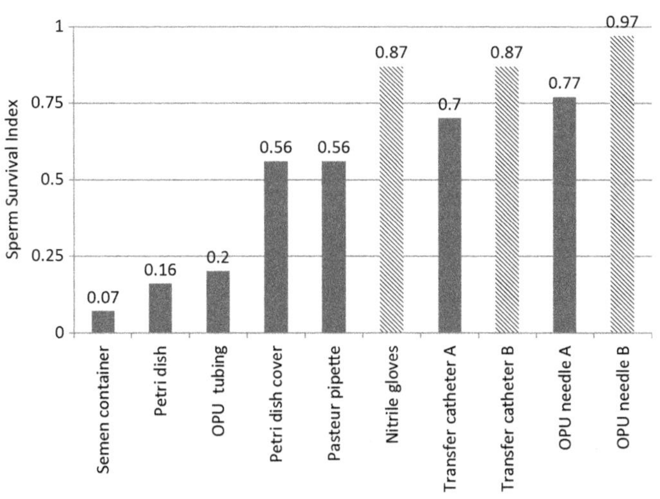

Fig. 2.3 Sperm viability index for a range of disposable supplies tested for sperm survival. Sperm viability index =% Final motility/% Initial motility. The values plotted represent the lowest result obtained in at least three different assays for the same item. Light bars represent items which passed the test with sperm viability index >0.85. *(Quartec GmbH.)*

Monitoring KPIs

Embryo development is a very sensitive QC indicator: in routine IVF culture a normal fertilization rate in the order of >65% and cleavage rate of >90% are expected (Hughes, 2012). The cleavage rate is important, as a block at the 2PN stage indicates a serious problem. At least 65% of inseminated oocytes normally result in cleaved embryos on Day 2 (see Fig. 2.4a, b). The appearance of the blastomeres and the presence of fragmentation are also useful indicators: healthy blastomeres in the early human embryo have granular cytoplasm of consistent appearance.

An ongoing record of the results of KPIs such as the rates of fertilization, cleavage, and embryo development provide the best short-term evidence of good QC. Graphic analysis of these parameters can reveal problems early so that corrective action can be taken promptly, minimizing the extent of any problem that might arise (Kastrop, 2003).

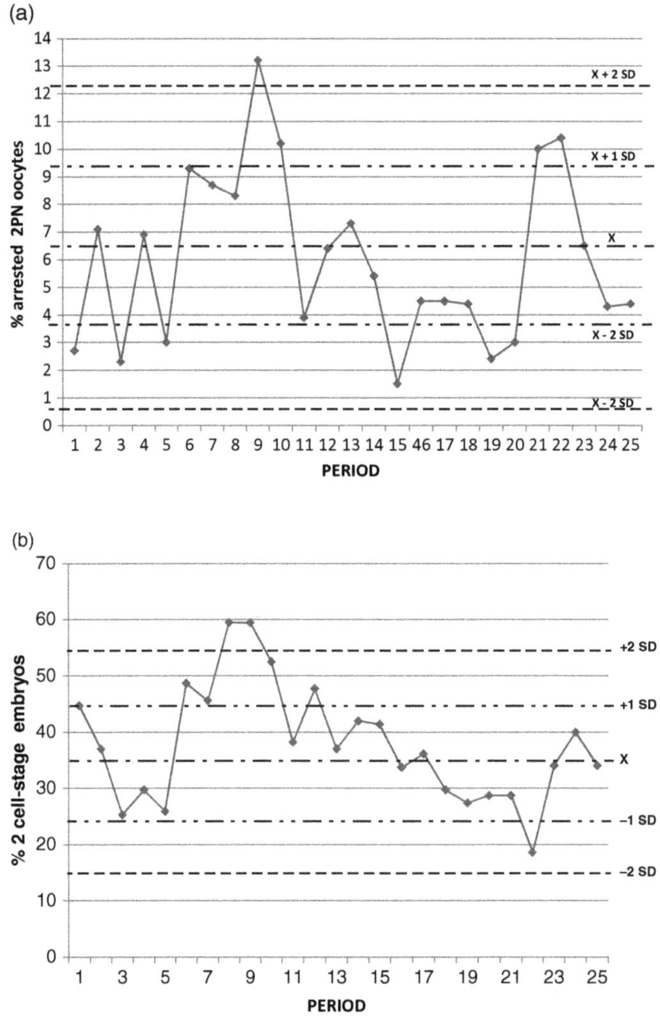

Fig 2.4 Examples of Levey–Jennings Quality Control Charts reporting results over 25 consecutive periods of 20 treatment cycles. From period 8 onwards, all procedures were reviewed with a return to normal after period 10. Intervals 21–22 again raised concerns leading to another general review of equipment and procedures.
(a) Percentage of arrested 2 PN oocytes. (b) Percentage of cleaved embryos. *(Quartec GmbH.)*

Graphs of KPIs can be plotted and compared with established limits for the mean, standard deviation (SD), and standard error of the mean (SEM), so that deviations can be easily recognized as warning or action points. Fluctuations in KPIs can be readily seen by plotting QC data on a graphical representation such as a Levey–Jennings chart (see Fig 2.4). These charts give a visual indication of whether or not the measured KPI is within acceptable limits. The date and time of each KPI measurement is plotted along the *x*-axis, and the measured KPI values entered on the *y*-axis. The chart is marked with lines that indicate mean KPI values, and one, two, and three SD limits on either side of the mean. Examining the pattern of points entered provides a simple way to detect random error, shifts, or trends.

In practice, applying this simple method to ART results presents the practical problem of defining the values for the mean and SD, e.g. for the mean and SD of pregnancy rate per embryo transfer. Large laboratories that perform thousands of treatments annually are able to calculate their own values, as well as compile separate Levey–Jennings charts according to relevant groups, such as female age or infertility type. Smaller laboratories may decide to use values that are based upon data provided by their national IVF register or an international survey program such as the International Committee Monitoring Assisted Reproductive Technologies (ICMART, www.icmartivf.org). However, the calculation of a mean based upon a small set of values introduces problems related to Type 1 and Type II errors, and corrections may reject the Null Hypothesis. In simple terms this means that it may be difficult to distinguish between false and true deviations for a single KPI. This can be resolved by selecting mean KPI values based upon a larger number of observations from a well-recognized data set, such as provided by *ICMART*.

A second difficulty lies in correct interpretation of those values that fall outside of two SDs from the mean: are they simply fluctuations, or significant changes that warrant improvement in quality? The "Westgard Rules" (http://www.westgard.com/) suggest using a combination of decision criteria based upon five different control rules (Multirule QC) in order to circumvent this dilemma. Each individual laboratory can define their own criteria for applying the rules, and how strictly they should be followed. A proposed model for detecting real drops in KPIs suggests that a fall in the value of a KPI below 2 SDs may indicate a problem, and may be taken as a warning that detailed survey is warranted. The following "rules" may be applied to determine that a real problem exists:

1. The value of a KPI is below 3 SDs of the mean.
2. Two consecutive values of a KPI fall below 2 SDs of the mean.
3. Four consecutive values of the KPI are below 1 SD of the mean.
4. Seven consecutive decreasing values of the KPI; some of the values may be within the range of mean ± 1 SD.

This proposed model does not exclude the possibility of false interpretation of a KPI drop; in addition, since treatment outcome may not be available for several weeks/months, the impact of a non-detected problem in the laboratory may potentially affect outcome for a significant period of time. It is therefore prudent to establish laboratory KPIs that might be used as an early warning system, to enable more immediate detection of something that is starting to go wrong in the gamete/embryo culture system. For example:

1. Have the number of normally fertilized oocytes completed the first cell division by 26 hours post-insemination or sperm injection?
2. Is the number of arrested zygotes low, and the ratio of two-cell embryos that cleave to four-cell embryos relatively high?

Both parameters can be calculated as a percentage for every 20 consecutive IVF or ICSI cycles. Since they represent a larger dataset, and are related to phenomena that occur during culture and before the final outcome of the treatment is known, these parameters are more appropriate KPIs for monitoring laboratory QC. In practice, retrospective analysis using Levey–Jennings charts can be used to plot the above data and calculate the corresponding implantation rates. Instead of using SDs of the mean as cut-off levels, critical values can be based on good implantation rates as reported in the literature. It is then possible to see whether there are any falls in these KPIs (i.e. arrested zygotes, low % four-cell stage embryos on Day 2, etc.) and relate these to any corresponding implantation rates that may fall below the critical value. This allows evaluation of the situation as a true positive, which can be used as early indication of a potential problem in the culture system. The sensitivity and the specificity of this KPI can then be considered and the efficiency gauged when used prospectively. Any trend or shift in data can be noted and investigated, with consideration of patient age and activity level: examine records for batches of materials (plasticware and media), air quality, temperature/CO_2 levels in incubators and flow cabinets, water quality, oil, and gloves.

In the UK, the HFEA has chosen to use **cu**mulative **sum** analysis to assess IVF clinics' performance in terms of success rates and multiple pregnancy rates. This is known as a CUSUM plot, derived from a statistical tool (statistical process control, SPC). SPC has become an increasingly effective means of assessing surgical and medical outcomes (Leandro *et al.*, 2005). When treatment cycles are reported to the HFEA, a CUSUM plot shows the difference between an individual clinic's success rate relative to the average success rate for all other registered clinics at that time. Plots can then be used to allow trends to be identified, e.g. if a clinic has consistently better results than the national average the plot shows a progressive rise, whereas consistently poor results show a progressive fall. Upper and lower alert limits are provided on the plots. If a clinic falls below the lower alert limit, the HFEA issues a warning to the clinic, so that their practice can be reviewed and changes implemented in order to try to improve performance.

Certification and accreditation

Recognition of a laboratory's competencies and quality systems by an ISO standard has concrete benefits:

1. Improvement in quality of care provided.
2. Recognition of individual training and competencies by providing education.
3. Improved management structure.
4. Cost reduction via quality improvement policies such as "Genba Kaizen," a Japanese concept of continuous improvement designed to enhance processes and reduce waste (Imai, 1997).
5. Patient satisfaction.

Box 2.3 Case report

IVF laboratories associated with University hospitals often have studies or projects running parallel to IVF clinical activities. Such projects may not only add an extra burden to the lab workload, but may also introduce a risk of mistakes with far-reaching consequences, as in the example below.

Material that cannot be used for treatment is often seen after IVF/ICSI, such as immature, activated (one PN), and polyploid oocytes. These cells provide a very valuable resource for research, but should be handled with specific delineated protocols to ensure that they are kept separate from the patients' clinical material, in a clearly labeled dish and in an incubator designated only for research.

Polyploid zygotes collected in an academic hospital setting for a metabolic study were kept in a routine lab incubator, and the resulting embryos were transferred to a different patient. The mistake was recognized immediately after the ET procedure, the couple was informed, and the embryos were flushed out of the uterus. The correct embryos were cryopreserved for transfer in a subsequent HRT cycle. The couple intended to sue the clinic, but abandoned the suit in return for financial compensation from the hospital. When the incident was investigated, it was found that double witnessing was carried out only after ET, and not before or during the procedure. The entire witnessing procedure was reviewed, and all forms adapted to include more detailed records of the double witnessing procedure.

The couple went to another clinic for treatment, but did not achieve a pregnancy. They returned to the original clinic, stating the following points:

- "Humans do make errors, but at least you honestly admitted the error and immediately informed the parties concerned.
- Your policy is transparent, with nothing hidden and no lying.
- We never felt that we were being treated as a 'number,' as we did in the other clinic.
- We were informed that you did review and amend your procedure."

This case report emphasizes valuable points:

1. Study protocols should be as clearly defined as any other SOP.
2. Clearly separate clinical and scientific activities, and conduct the scientific activities in a different laboratory if possible.
3. Perform internal audits regularly to confirm that all procedures are executed as written in the SOP.
4. Review SOPs on a regular basis, testing them for validity; use only approved versions.

6. Third-party recognition for reimbursement of treatments.
7. Better market position.

Before initiating a process towards certification/accreditation, it is important to understand that these represent two separate entities.

Certification: a neutral independent third party confirms (certifies) that a system complies with an internationally accepted standard, such as an "ISO Norm." In practice, this means that laboratories can search for a company that will assess whether or not their quality system has the structure required by the ISO Norm. A wide and varied choice of companies offers this service, and their fees may differ considerably. Certification provides testimony that a correctly structured QM is in place, according to an internationally accepted standard. Certification is often presented to patients as reassurance and guarantee of the clinic's high quality, but this is not the case. These certificates merely confirm that there is a quality management whose structure corresponds to an internationally recognized ISO standard: the appropriate structure is in place, but there is no guarantee of the content. For example, an IVF laboratory with very poor results can be certified, but will never be accredited because poor treatment outcomes usually indicate a lack of competence.

Accreditation requires evidence that your laboratory has the appropriate competencies for the service provided, including adequate facilities/space, equipment/instruments, and personnel who are adequately and continuously trained and able to demonstrate the competence to fulfil their tasks. *Accreditation can only be granted by a "National Accreditation Office,"* and there is no elective choice in selecting the third party who will audit the clinic/laboratory. If the laboratory seeks accreditation in a very specific field, e.g. "comparative genome hybridization," a local/national expert might not be available, and the national accreditation office might appeal for an expert from a neighboring country. The national/federal accreditation office is likely to be a member of the International Association for Laboratory Accreditation ILAC (International Laboratory Accreditation Corporation, https://www.ilac.org/members_contact_details.html) or EA (European co-operation for Accreditation, http://www.european-accreditation.org).

It is important to note that certification and accreditation are valid only for a limited period (2 to 5 years), and re-certification and surveillance are required in order to maintain validity of the acquired certification/accreditation. Nicely framed certificates on display at the clinic entrance are completely meaningless if the date is more than 2 to 5 years ago, respectively.

The Accreditation process involves a series of logical steps:

1. Purchase a version of the required ISO standards: http://www.iso.org.
2. Request written procedures from all staff involved, describing who is responsible for each procedure, how it is carried out and what instruments/supplies are required. Use a standardized form to produce SOPs.
3. Analyze interactions between different sections of each department and construct flowcharts to define the sequence and the inter-relationships between procedures.
4. Compile a QC manual.
5. Adapt facilities as appropriate/required.
6. Establish a QM that complies with the requirements of ISO9000:2015 certification.
7. Ensure that all calibration certificates contain the information required, i.e.:
    a. Identification and accreditation numbers, the ISO Norm (usually ISO17025), and name and address of the company responsible for the calibration.
    b. Parameters (room temperature, humidity, etc.) that are monitored in order to achieve accurate calibration.
    c. Method used for calibration.
    d. Reference standards used for calibration and their traceability.
    e. Calibration conditions, i.e. the values of parameters under (b).
    f. Measured values.
    g. Results of calibration and margin of error (degree of uncertainty). If the margin of error is too great, the relevance of measured values is questionable, and may not allow differences to be perceived. The margin of error may sometimes be great simply because no other calibration method is available.

  h. Date of previous calibration, and the date that the next is due.
  i. Name and signature of the technician responsible.

(NB: Correctly documented calibration certificates are uncommon.)

8. Submit an application for accreditation; this is followed by a preliminary visit that includes a discussion with the national/federal representative before proceeding to the next step. The representative will confirm that you have selected the correct accreditation standard with respect to the field, and the competencies required for the services that you provide (there is only one standard for medical laboratories, ISO15189).

The ISO15189 Accreditation Norm requires an established Quality System according to ISO9001 certification. A clinic can compile a checklist based on each clause of the ISO, and use this for pre-assessment audit. When all the requirements of the checklist have been fulfilled, you are ready for a first "preliminary audit."

Audit visits and non-compliances

An audit is defined as a detailed examination of an organization and the specific processes carried out within it, with inspection to evaluate whether:

1. An effective quality system is in place according to the ISO9001:2008 standard.
2. Measurements are valid and performed with properly calibrated instruments.
3. All procedures are effectively documented.
4. Audit systems and an audit schedule are in place, with concrete evidence of compliance with protocols.
5. Staff training is adequate according to appropriate standards, and a logbook is maintained with traceability to ensure that the minimal requirements (including theoretical aspects and comprehension, as well as skills) are fulfilled before any member of staff is allowed to perform a procedure independently (see Chapter 12).

A laboratory audit/inspection can potentially identify two types of non-compliance:

1. Those that need immediate corrective action and lead to immediate interruption of the services concerned.
2. Those that do not require immediate action, but the required corrective action must be carried out within a specified time period.

Preventive and corrective action

Preventive action prevents a problem from ever taking place, and corrective action takes place once a problem has occurred. According to ISO9000,

- *Preventive action* is "an action to eliminate the cause of a potential nonconformity or other undesirable potential situation." This action is proactive in the sense that opportunities for improvement are identified before a problem is identified.
- A problem that has been identified requires *corrective action*, i.e. reaction to determine and correct the immediate effect of the non-conformity. Corrective action removes the root cause of the problem in order to prevent further incidents/problems.

Documented procedures are required to deal with corrective and preventive action, as two separate and distinguishable events, clearly defining responsibilities for identifying, evaluating, and reviewing both types of action. Monitoring preventive and corrective action can take into account trend analyses of processes, KPIs, whether any alarms have been triggered, and risk analyses. Honest full reports of all adverse events and incidents can provide a traceable method for identifying where corrective and preventive action is effective. With this in mind, a "no blame" culture is beneficial to an IVF clinic. This means that errors can be tolerated, provided that the staff are honest and report any mistakes, and also learn from these mistakes. This empowers staff and encourages discussion of errors among work colleagues, preventing "cover ups" and encouraging an honest workforce where mistakes are identified and ultimately eliminated.

Auditors cannot advise on corrective action: this could potentially lead to conflict, should the action undertaken by the clinic turn out to be at a level that is unsatisfactory and still results in non-compliance. Good auditors may give advice or suggestions, but warnings cannot be issued for possible non-compliances in the future, and these should never appear in the final report. Remember that the auditor is not responsible for granting accreditation (or not): this will be granted by the national/federal office after they have examined the auditor's report. The auditor's task is to verify compliance, not to condemn or to approve.

A *horizontal* audit evaluates a given process or activity that takes place within an organization across more than one department. These audits are used to gauge the effectiveness of a particular process or activity relative to the same activity performed by other groups. This type of audit is also known as "peer audit."

A *vertical* audit focuses on a single procedure (rather than a process), e.g. semen preparation for IUI and how it is dealt with throughout the clinic: specimen collection, identification, and handling. It will include direct questions to operators, verification of related documentation, and details of instrument calibration. This type of audit is usually

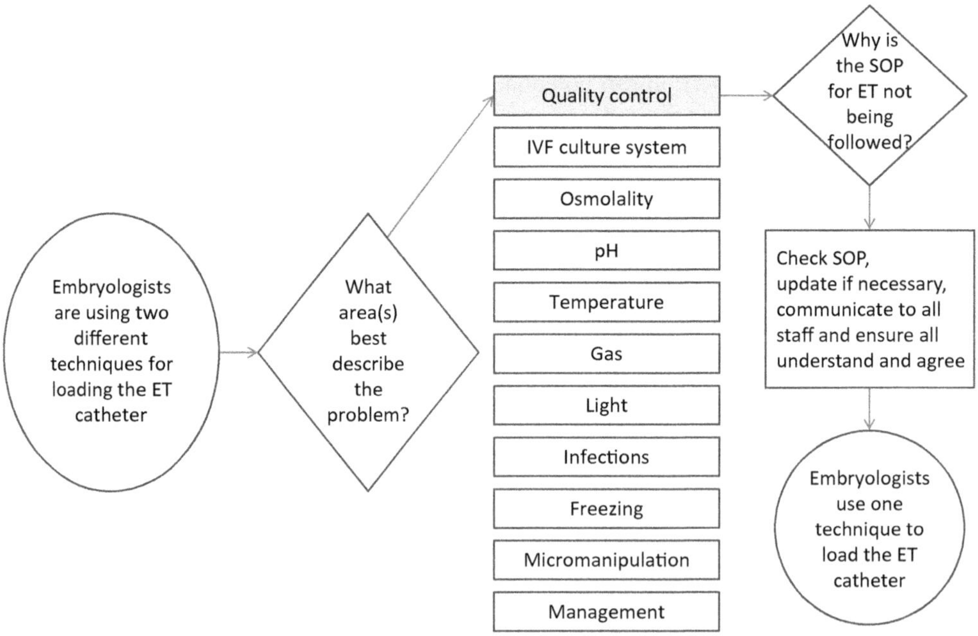

Fig. 2.5 Process map.

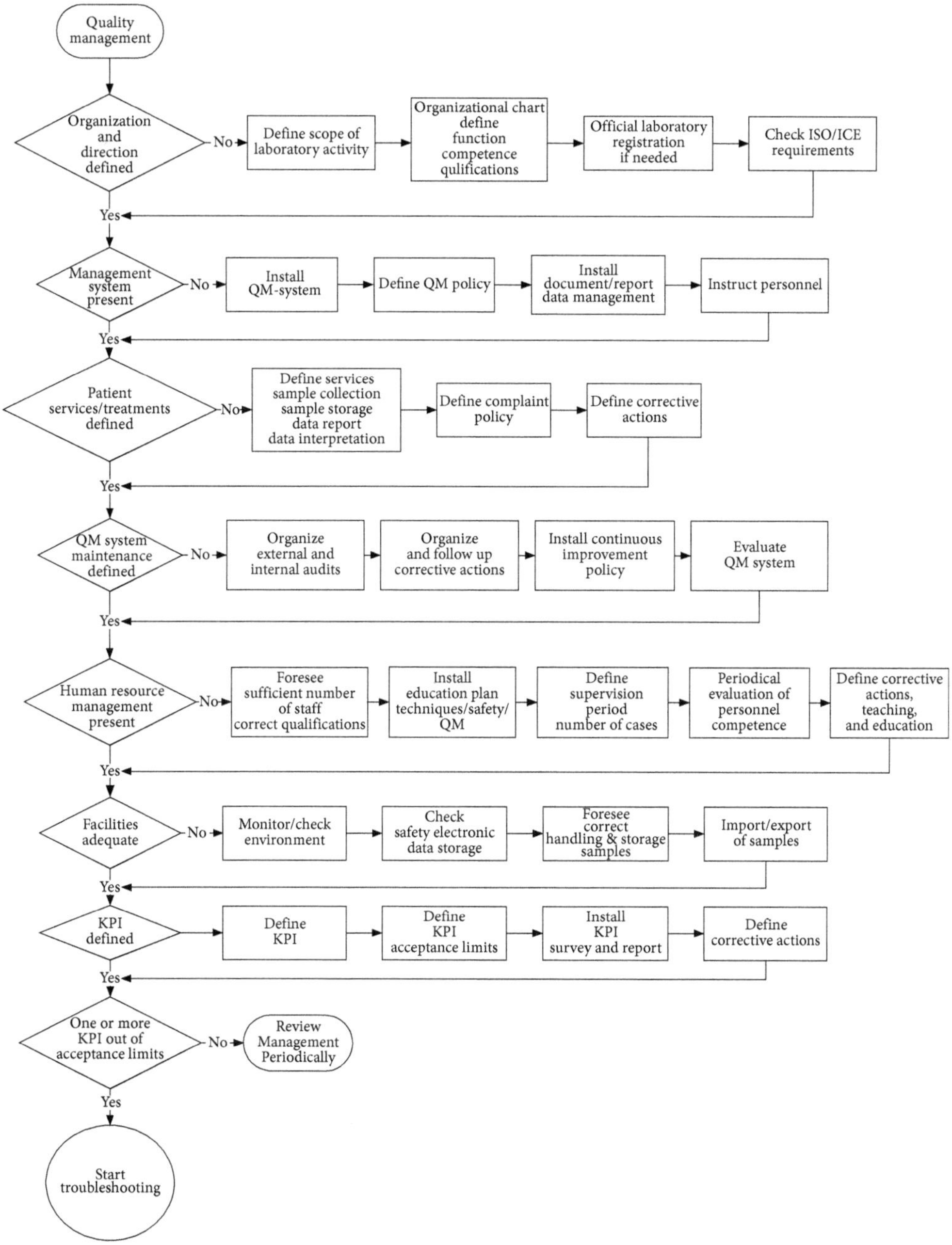

Fig. 2.6 Flowchart.

performed during surveillance audits; it requires full preliminary preparation on the part of the auditor, who will confirm that the correct procedures are in place. SOPs may (or may not) have been adapted according to corrective actions requested in a previous audit report.

A first audit visit is usually horizontal, and takes place according to well-established rules and time schedules to include a program, an introductory meeting, and a concluding meeting during which the laboratory will receive a list of non-compliances. Reserve a meeting room, and prepare all documentation according to the requirements of the audit program. If hospitality is provided, there is no need for gastronomic lunches (a good lunch will not deflect from non-compliances!).

Documents are supplied to the auditors in advance of the actual assessment visit. The audit usually begins by visiting the facilities after the introduction meeting. The inspection will include a detailed examination of all documents/documentation, interviews with selected staff (all staff members should be available including the consultants), and process monitoring. The findings are summarized during a final meeting, with brief discussion of the conclusions. All key personnel should be present at both the introductory and the final meetings. If several auditors are involved, e.g. a lead auditor and professional experts, they often meet separately during the day.

A laboratory audit is an impartial and objective assessment, not an "inquisition": it provides a unique opportunity to prove competence. You will be informed about the choice of expert who will visit, and are at liberty to disagree if there is any reason to doubt his/her impartiality, or potential conflicts of interest.

The process of obtaining certification/accreditation involves several months of preparation, and it is essential that the entire laboratory is convinced of the real benefit gained by the investment of time and effort. The process is often initially imposed by management "from above", and thus may be poorly accepted by the laboratory team, who perceive it to be a great deal of extra paperwork. Good quality management resulting in continuous improvement is a system that is initiated by the laboratory team as a quest for recognition of their daily efforts. This type of system is characterized by its simplicity and efficiency, with the ability to introduce corrective action and a minimum of paperwork (Figs. 2.5, 2.6).

Further reading

Association of Clinical Embryologists – www.embryologists.org.uk

ASRM Guidelines

The Practice Committee of the American Society for Reproductive Medicine and the Practice Committee of the Society for Assisted Reproductive Technology (2008). "Revised guidelines for human embryology and andrology laboratories." *Fertil Steril*, **90**: S45–S59.

Boone WR, Higdon HL III (2012). Quality control management. In: *Practical Manual of* In Vitro *Fertilization, Advanced Methods and Novel Devices*. ZP Nagy, AC Varghese, & A Agarwal (eds), Springer NY, pp. 33–9.

Carson B, Alper M, Keck C (2005). *Quality Management Systems for Assisted Reproductive Technology ISO 9001: 2000*. Taylor Francis.

College of American Pathologists, Accreditation and Laboratory Improvement http://www.cap.org/apps/cap.portal?_nfpb=true&_pageLabel=accreditation

Cutting R, Pritchard J, Clarke H, Martin K (2004). Establishing quality control in the new IVF laboratory. *Hum Fertil*, 7(2), 119–25.

Department of Health – www.dh.gov.uk.

Edwards RG, Purdy JM (1982). *Human Conception In Vitro; Proceedings of the First Bourn Hall Meeting*, RG Edwards & JM Purdy (eds.) Academic Press, London.

Elder K, Dale B, In Vitro *Fertilization*, 3rd edition; Cambridge University Press, 2010 Joint Commission International (JCI). http://www.jointcommissioninternational.org/.

Hreinsson J, Sjøblom P (2012). Regulation, licensing and accreditation of the ART laboratory in Europe. In: *Practical Manual of* In Vitro *Fertilization, Advanced Methods and Novel Devices*, ZP Nagy, AC Varghese, & A Agarwal (eds), Springer NY, pp. 611–18.

Hughes C (2012). *Association of Clinical Embryologists – Guidelines on Good Clinical*

Practice in Clinical Embryology Laboratories. Hum Fertil (Camb), **15** (4), 174–89.

International Organisation for Standardisation (ISO) (www.iso.org).

BA Keel, JV May, De Jonge CJ, (eds.) (2000). *Handbook of the Assisted Reproduction Laboratory*, CRC Press.

Lane M, Mitchell M, Cashman KS, Feil D, Wakefield S, Zander-Fox DL (2008). To QC or not to QC: the key to a consistent laboratory? *Reprod, Fertil, Devel*, **20**, 23–32.

Magli MC, Van den Abbeel E, Lundin K, Royere D, Cen der Elst J, Gianaroli L (2008). ESHRE Guidelines. Revised guidelines for good practice in IVF laboratories. *Hum Reprod* **23**(6) ESHRE pp. 1253–62.

Mortimer D, Mortimer S (2007). *Quality and Risk Management in the IVF Laboratory*; Cambridge University Press.

References

Bertheussen K, Holst N, Forsdahl F, Høie KE (1989). A new cell culture assay for quality control in IVF. *Hum Reprod*, **4**(5), 531–5.

Critchlow JD, Matson PL, Newman MC, Horne G, Troup SA, Lieberman BA (1989). Quality control in an in vitro fertilisation laboratory: use of human sperm survival studies. *Hum Reprod*, **4**, 545–9.

Elder K, Johnson MH (2015c). The Oldham Notebooks: an analysis of the development of IVF 1969–1978. III. Variations in procedures. *Reprod Biomed and Society*, in press.

George MA, Braude PR, Johnson MH, Sweetnam DG (1989). Quality control in the IVF laboratory: *in-vitro* and *in-vivo* development of mouse embryos is unaffected by the quality of water used in culture media. *Hum Reprod*, **4**, 826–31.

Gilligan A, Schimmel T, Esposito B Jr., Cohen J (2008). Release of volatile organic compounds such as styrene by sterile petri dishes and flasks used for in-vitro fertilization. IVF Online newsletter article: http://www.ivfonline.com/User/Newsstand/Release.aspx.

Hall J, Gillian A, Schimmel T, Cecchi M, Cohen J (1998). The origin, effects and control of air pollution in laboratories used for human embryo culture. *Hum Reprod*, **13** (Suppl 4), 146–55.

Holyoak GR, Wang S, Liu S, Bunch TD (1996). Toxic effects of ethylene oxide residues on bovine embryos *in vitro*. *Toxicology*, **108**, 33–8.

Imai M (1997). *Genba Kaizen: A Commonsense Low-cost Approach to Management.* New York: McGraw-Hill Professional. p. 13.

ISO/IEC 17025 (2005). General requirements for the competence of testing and calibration laboratories.

ISO 9001 (2008). Quality Management Systems.

ISO15189 (2012). Medical Laboratories – Requirements for Quality and Competence.

Johnson MH, Franklin SB, Cottingham M, Hopwood N (2010). Why the Medical Research Council refused Robert Edwards and Patrick Steptoe support for research on human conception in 1971. *Hum Reprod*, **25** (9), 2157–74.

Kastrop PM (2003). Quality management in the ART laboratory. *Reprod Biomed Online*, **7**(6), 691–4.

Leandro G, Rolando N, Gallus G, Rolles K, Burroughs AK (2005). Monitoring surgical and medical outcomes: the Bernoulli cumulative SUM chart. A novel application to assess clinical interventions. *Postgrad Med J*, **81**, 647–52.

Nijs M, Franssen K, Cox A, Wissmann D, Ruis H, Ombelet W (2009). Reprotoxicity of intrauterine insemination and *in vitro* fertilization – embryo transfer disposables and products: a 4-year survey. *Fertil Steril*, **92**(2), 527–35.

Purdy JM (1982). Methods for fertilization and embryo culture *in vitro*. In: *Human Conception In Vitro; Proceedings of the First Bourn Hall Meeting*, RG Edwards & JM Purdy (eds.), Academic Press, London, p.135.

Rinehart JS, Bavister BD, Gerrity M (1988). Quality control in the in vitro fertilization laboratory: comparison of bioassay systems for water quality. *J In Vitro Fertili Embryo Transfer*, **5**, 335–425.

Van den Bergh M, Baszo I, Biramane J *et al.* (1996). Quality control in IVF with mouse bioassay: a four years experience. *J Assist Reprod Genet*, **13**, 733–8.

Chapter 3

The IVF culture system

Introduction

Gametes and embryos are maintained in culture systems that aim to mimic *in vivo* conditions, with buffered media that provide a stable environment via complex interplays between physico-chemical parameters of pH, temperature, and osmolality. These three parameters are dependent upon, and are directly influenced by, the composition of the media and its buffering systems, as well as by the gas atmosphere and incubation conditions.

Optimizing media is a perfect example of troubleshooting a complex system that is full of poorly understood interactions. During the 1950s, Whitten, one of the early pioneers of embryo culture, solved the problem of calcium precipitation in the presence of sulfate anions by replacing $CaCl_2$ with calcium lactate; this also circumvented the problem of accurately weighing the correct amount of $CaCl_2$, which is very hygroscopic. Coincidentally, this strategy also demonstrated that lactate could function as an exogenous energy source in the development of mouse two-cell embryos to the blastocyst stage (Whitten, 1956). Irreversible precipitation of calcium as $CaCO_3$ and calcium phosphate in aged media can still be observed, particularly if it has attained a very alkaline pH, or has been stored in a refrigerator that is too cold (see Fig. 3.2). Despite a great deal of input from expert culture media pioneers (Whitten, 1956; Biggers *et al.*, 1957; Brinster, 1963; Chang, 1967; Wales & Whittingham, 1967, 1973; Whitten & Biggers, 1968), for several decades culture media remained nothing more than buffered salt solutions supplemented with lactate, pyruvate, and glucose as energy sources. A classic example is Earle's Balanced Salt Solution (EBSS), a recipe still used today as a basis for many commercial media used in human IVF. Biggers applied the "simplex optimization method," a multifactorial analytical approach that led to the development of a medium with higher potassium (K) concentration, KSOM (Potassium Simplex Optimized Medium, Lawitts & Biggers, 1991; Summers, 2014). More complex media, such as Ménézo's "B2" medium, were developed during the 1970s (Ménézo, 1976). Quinn and colleagues developed media based upon the composition of human tubal fluid (HTF, Quinn *et al.*, 1985) and media formulations have continued to evolve over the past four decades, as comprehensively described by Quinn (Quinn, 2012, 2014).

Familiarity with all basic principles of biochemistry and physiology were fundamental to creating the field of human IVF (Elder & Johnson, 2015c), and still remains the essential starting point for troubleshooting an IVF culture system. This has become increasingly important in contemporary IVF, which relies upon supplies from numerous different commercial sources. Case studies clearly demonstrate that nothing should be taken for granted when dealing with human IVF culture systems; this chapter outlines a brief summary of the fundamental principles that underpin successful human IVF culture. A complete and up-to-date review is provided in *Culture Media, Solutions and Systems in Human ART*, ed. Patrick Quinn, 2014.

Culture media

The specific needs of embryos during pre-implantation development have by no means been completely defined. Data regarding human embryo metabolism started to become available nearly a decade after the birth of Louise Brown (Leese *et al.*, 1987), and advances in technology continue to bring new insights to the complexity of the systems involved (Ménézo *et al.*, 2013). Continuing research has led to numerous changes in media formulations, each with their advantages and disadvantages, as reviewed by Biggers & Summers (2008) and by Quinn (2012). Media formulations have become more complex; the issue of ongoing media development and its future was recently reviewed (see "The future of human embryo culture media – or have we reached the ceiling?" Zander-Fox & Lane, 2012). The possibility that an embryo can be "improved" is highly unlikely, but an optimal culture system should minimize the incidence of non-viable embryos by eliminating all sources of potential stress that jeopardize gamete and embryonic health. The period before embryonic gene activation is the most vulnerable, as development is entirely dependent upon mRNA stored in the oocyte, and mechanisms to repair damage are very limited (Ménézo *et al.*, 2013). Further development involves signal diffusion, either via extracellular diffusion or via intercellular junctions at later stages (Dale *et al.*, 1991), and both of these mechanisms are highly sensitive to any fluctuations in culture conditions. Increasing media complexity has been accompanied by concerns about long-term safety (Campbell *et al.*, 2011). An early example was the observation of large offspring syndrome in cattle and sheep following the transfer of embryos that had been cultured in medium supplemented with serum (Young, 1988). Data collected over the past two decades by national IVF registers suggest that *in vitro* cultured human embryos can be vulnerable to media composition, both in the short and the long term (Velker *et al.*, 2010). Several studies report that culture conditions have a direct impact on embryonic transcription and translation (Ho *et al.*, 1994; Wrenzycki *et al.*, 1999, 2001; Khosla *et al.*, 2001); on the health of babies born (Vergouw *et al.*, 2012); and on embryonic metabolism in general, such as acceleration of protein turnover and impairment of mitochondrial function (Jung, 1989).

Spontaneous generation of reactive oxygen species (ROS) in buffers and enriched media has been described (Cohen, 1987; Michelson, 2000; Grzelak *et al.*, 2001). Martin-Romero and co-workers (2008) confirmed that the generation of ROS in commercially supplied culture media and buffers varies according to their composition. Free oxygen radicals lead to generation of hydrogen peroxide (H_2O_2), especially in the presence of HEPES buffer or other amines in the medium. H_2O_2 can react with oxygen free radicals, starting a chain reaction that amplifies the production of ROS. Subsequent potential effects include dysregulation of calcium balance, which plays a critical role in signalling pathways during fertilization (Dale & DeFelice, 1990).

ROS also deplete reserves of glutathione (GSH), the most abundant intracellular antioxidant, and this promotes the formation of disulfide bonds in redox sensitive proteins (Cumming *et al.*, 2004). GSH is a substrate for glutathione-5-transferase, which catalyzes detoxifying conjugation reactions, and for glutathione peroxidase, which protects the cell membrane from peroxide damage that can result in cell lysis. ROS production in media can be diminished by the addition of a metal ion chelator (ethylenediaminetetraacetic acid, EDTA), which inhibits enzyme-driven reactions in ROS-generating pathways that need metal cofactors. Follicular fluid and phosphate buffered saline (PBS) do not produce significant amounts of ROS, and adjusting oocyte incubation times with this in mind might be advantageous (Guérin *et al.*, 2001; Martin-Romero *et al.*, 2008). Suboptimal culture conditions can also jeopardize embryo viability by influencing the process of imprinting (Hiura *et al.*, 2012).

Significant problems with media composition or inappropriate culture conditions may result in immediate cleavage arrest or cell lysis. However, it is still possible that embryo culture might be successfully maintained *in vitro*, with apparently good morphological characteristics, and less obvious problems related to embryo viability might not be apparent until the time of birth or after extended follow-up. For this reason, it is important for every laboratory to keep meticulous embryo culture records, and disclosure of the exact medium composition should be a legal obligation for companies that provide commercial media.

When selecting culture media for use in an IVF system, it is useful to consider availability of information about:

- Composition of each batch (not always easy to obtain)
- Quality control during production
- Endotoxin testing
- Evidence of use in human IVF – i.e. proven track record
- Certificates of analysis (be aware that in Europe the CE label indicates only conformity to European manufacturing guidelines, and is *not* a guarantee of quality)
- Serum or serum derivatives from virally screened sources
- Use of appropriate bioassay for testing as defined by national and international regulations.

Although some companies provide information regarding mouse embryo assay (MEA) testing, the benefits of these assays are questionable, since media can pass a MEA, but still be toxic to human embryos. Oil used for culture overlays is a particularly good example: case studies of serious problems that were found to be due to commercially supplied (MEA-tested) oil are numerous! Often the toxicity is not apparent for embryos grown to Day 2 or Day 3, but becomes apparent after extended culture to blastocyst stage, when significantly fewer embryos survive to Day 5. An in-house sperm survival toxicity test is recommended for all batches of media and oil, in order to ensure optimal function in the individual clinic's culture system (see Chapter 2).

Keep records of every delivery, with documented details of:

- Batch number (note: for some European companies, the batch number on the bottle may differ from the batch number recorded on the box/container)
- First date of use for each batch (note the date of bottle opening)
- Key Performance Indicators (KPIs) relevant to the use of each batch.

Practical guidelines for handling of media

1. Ensure that the cold chain is not broken during transport of culture media between manufacturer, supplier, and clinic. This can be verified by asking the provider to add a

Box 3.1 Case report

A shipment of media arrived at the IVF laboratory in a cold-storage container ("esky") that was broken, and the media was no longer cold when it was unpacked. Investigation revealed that the damage had apparently been caused by the delivery courier; the media company did not accept responsibility for the damage (Fig. 3.1).

Fig. 3.1 Broken cold-storage container ("esky") used for media shipment.

small USB temperature data-logger to the inside of the transport box. These are small, relatively cheap, and although not extremely accurate, can still reveal any significant fluctuations in temperature during shipping. Bear in mind that the company producing the media may not be the one to manage the shipping, and that custom controls may introduce considerable time delays during transport in some countries. The choice of culture medium should therefore be based not only on costs and composition, but also on the availability and the quality of shipping (Box 3.1).

2. Check the production schedules of the media and place your order as close as possible to their production dates, so that media arrives with the maximum shelf-life.
3. Avoid exposure to light: light has wave and particle characteristics, and photons must be considered as small energy packets able to generate free radicals (Pomeroy & Reed, 2013; see Chapter 7). Where media is supplied in boxes, keep the bottle in this packaging so that the contents are not continuously exposed to light, especially if it is stored in a refrigerator with transparent doors.
4. Store the media continuously at the appropriate temperature. Some laboratories have only "kitchen refrigerators" available, which are generally too humid and are prone to

large temperature fluctuations. Refrigerators designed for medical storage with transparent gliding doors are recommended: being able to see what you need to remove allows partial opening of the sliding door, and this will limit temperature variations within the refrigerator. Continuous temperature monitoring is useful, and this is included in regulations such as ISO15189 and the EU Tissues and Cells Directive.

5. Media is sometimes aliquoted into smaller volumes after delivery, so that each aliquot can be used for a single patient's gamete preparation and culture. Check the smaller aliquot containers to make sure that they are appropriate for culture media storage: cell culture plasticware may release considerable amounts of volatile organic compounds (VOCs) (see Chapter 8), and off-gassing for several days is recommended. Bisphenol A (BPA), used in the manufacture of many plastic products, is known to have significantly harmful effects on reproductive function (Mok-Lin *et al.*, 2010), and causes oocyte aneuploidy in experimental animals. Experiments in a mouse system also confirmed that exposure to BPA at early developmental stages has a stable effect on imprinting, and maternal dietary supplements with methyl donors (folic acid) or genistein (present at high levels in soya) could counteract this effect (Dolinoy *et al.*, 2007).

   Carry out aliquoting under sterile conditions, being aware that every contact with a pipette or any other laboratory equipment may introduce embryotoxic substances. Some commercial companies may advise against aliquoting, as this invalidates the quality assurance (QA) for the media taken directly from the bottle provided. Freezing aliquots of culture media further invalidates QA and is not recommended. Companies package media in different volumes: good laboratory management should dictate that media is ordered in the appropriate quantities and aliquots for daily usage, so that sufficient, but not too much, media is supplied. Place orders for quantities that are optimal to ensure safety and quality.

6. Careful attention to "shelf-life" is a critical factor, and is dependent on specific composition and conditions of storage: generally, the more "complex" the media, the shorter its useful lifespan, with an average of approximately 4–6 weeks. If the media contains phenol red, the company may pre-gas the media prior to dispatch. This is essentially a visual esthetic for the consumer, as all media must be properly equilibrated prior to use (Box 3.2).

7. Media should be equilibrated for temperature and pH for 4–6 hrs before use, or by overnight incubation; however, the composition of the media must be taken into consideration, since incubation in certain types of media has been found to trigger a depletion in oocyte GSH content after 6–8 hrs (Martin-Romero *et al.*, 2008).

Box 3.2 Case report

A commercial supplier of culture media withdrew the pre-gas phase before delivery, so that on arrival in the IVF laboratory the appearance of the media (very pink!) caused major concern among some embryologists. The company received so many complaints that representatives were sent to Europe to hold a special seminar in order to reassure clients that the manufacturing was still tightly controlled, and that the media was still subject to tight QA. However, after publicly providing this reassurance, the company decided to re-introduce the pre-gassing phase after production, restoring the "expected" neutral pH color.

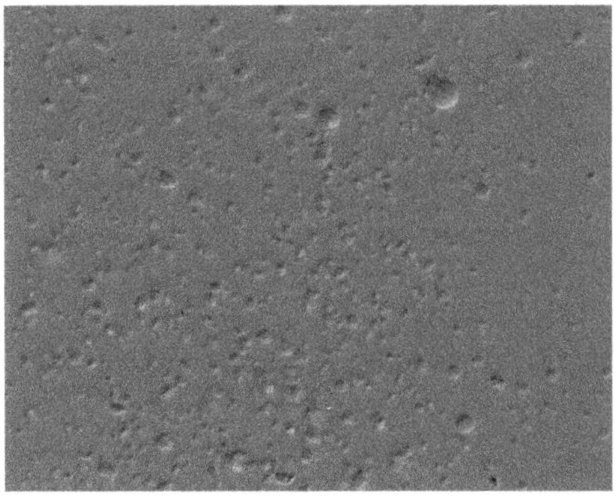

Fig. 3.2 Example of a precipitate (probably calcium phosphate) commonly observed in media, possibly as a result of being kept at a temperature that was too low, or due to frequent opening of the container. (*Image kindly provided by Dr. N Çiray, University of Leeds.*)

This depletion was not observed with incubation times shorter than 2 hours. For pre-gassed media, the desired pH may be reached sooner. Once a bottle has been opened, bicarbonate will dissipate, and partial removal of medium increases the head space in the bottle: media from pre-opened bottles may therefore take longer to equilibrate. The equilibration time required in order to reach the desired temperature and pH should be confirmed by a validation procedure.

8. Temperature and pH fluctuations during storage and handling may cause the following problems:
   - $CaCO_3$ may irreversibly precipitate if a flask is exposed to normal atmospheric conditions for too long, indicating that the media has become alkaline (see Fig. 3.2).
   - Precipitates of calcium phosphate may form if media is kept at a temperature that is too low.
   - Pyruvate solutions are stable for up to 24 months at 2 °C to 8 °C. However, in acidic conditions pyruvic acid will tend to polymerize and decompose.
   - Glutamine decomposition produces ammonium ions (NH_4^+); at 37 °C, glutamine decomposes by 10% per day, whereas at 4 °C decomposition of 10% would take 9 days (Tritsch, 1962). These rates were calculated using the same energy of activation (11.6 kcal/mole) for glutamine in a buffer at pH 7.2. Production of ammonia (NH_3) has been a topic of ongoing debate since Lane & Gardner (1994) attributed the birth of anencephalic mice to NH_3 in the culture medium. Levels of NH_3 (around 150 mM) in medium containing glutamine (1 mM) incubated at 37 °C for 24 hours were significantly higher than levels that were shown to inhibit embryo development (Lane, 2003). However, these results have been disputed as causing undue concern: the formation of NH_3 does not relate to glutamine depletion in a linear fashion, and glutamine can form the cyclic compound pyroglutamic acid (Ménézo & Guerin, 2007). Biggers (2004a) also questioned the use of ammonium sulfate, and designed media containing glycyl-L-glutamine, which is more stable than glutamine (Biggers, 2004b). Furthermore, Anckaert *et al.* (2009) confirmed that aberrant imprinting at regulatory sequences in mouse oocytes was not induced by NH_3 accumulation in a routine IVF culture system. Finally, it should be noted that cumulus cells convert pyruvate to alanine, and this may have a detoxifying function by removing endogenous NH_3; the presence of cumulus cells may improve oocyte quality in some culture systems (Ebner *et al.*, 2006).

Media for dish and tube preparation are supplied with QA if used directly from the bottle. Gassing bottles of medium in-house with a 5%–6% CO_2 mixture after opening invalidates the QA provided by the media company. If gassing is still preferred, the following points should be noted:

- Blowing gas on a liquid will increase evaporation and thus increase osmolality; it also increases the O_2 content and decreases protection against ROS.
- Albumin has a hydrophobic core, with hydrophilic side chains, and this gives the molecule detergent properties. Blowing gas through media containing albumin or another protein will cause the liquid to foam; this is not only impractical, but may contribute to protein denaturation.

Media conditioning

Regulatory authorities now oblige most laboratories to use "ready to use" media; "home-made" media, although still commonplace in animal IVF, have almost disappeared from clinical practice in countries with IVF legislation/regulation. Media formulations are basically buffered salt solutions that must be conditioned in order to become functional in culture. "Conditioning" involves adjustment of osmolality and pH under appropriate temperature regulation, supplementation with a protein macromolecule, and supplementation with pyruvate and glucose as energy sources and amino acids to sustain prolonged culture.

Protein supplement

In the early days of IVF, media were supplemented with 10% heat-inactivated patient serum, or fetal cord blood serum (if the laboratory was close to the maternity department). The percentage of serum was increased to 70% for embryo transfer, as it was thought that the increased viscosity might assist the transferred embryos to implant, or at least stay in the vicinity of deposition after transfer. With increasing numbers of cycles, the logistics of collecting and preparing individual maternal serum became a serious burden, and some clinics began to use pooled sera from a donor, sometimes taken from a fertile female, aged 20–35 years, who had just ovulated (often a member of staff!) (Box 3.3).

Albumin solutions available from blood banks were also commonly used in culture media (Ménézo, 1976); however, some preparations contained high concentrations of mould and bacterial growth inhibitors such as mandalic and proprionic acid, with unknown effects on embryo culture. By law, human serum is now prepared by heating a 20% solution of albumin in NaCl with caprylic acid and acetyl tryptophan at a temperature >45°C for 4 hours.

Box 3.3 Case report

During 1991, the use of pooled donor sera resulted in an episode of hepatitis B cross-contamination: 24 women conceived following IVF treatment, but suffered acute hepatitis B infection. Five pregnancies aborted spontaneously, and the remaining 19 pregnancies led to the birth of 24 children (van Os *et al.*, 1991). Although the authors stated that there was no evidence of any harmful effects in the newborn due to exposure to hepatitis B virus in the embryonic or fetal period, the episode confirmed that the use of poorly screened human serum in IVF is unacceptable.

Albumin is the major protein constituent of the embryo's environment; it can be incorporated by the embryo, binds lipids, and may help to bind and stabilize growth factors. Fatty acids (FAs) are an early requirement in embryo metabolism, required for membrane synthesis. FAs are mainly bound to albumin, but some albumins are heavily contaminated with lipid peroxides, which can be highly embryotoxic (Hammamah & Ménézo, 1999). Theoretically, albumin may not be indispensable, but it can effectively replace serum, and has a major role both in maintaining embryo viability and in preventing gametes and embryos from "sticking" to glass or plastic surfaces, facilitating their manipulation. Preparation of sperm in media without protein causes the sperm to stick to the sides of the test tube. Working without albumin has been attempted (Ali *et al.*, 2000, 2004), but protein, or an equivalent macromolecule (e.g. polyvinyl alcohol or hyaluronate) is required in human IVF for several valid reasons (Blake *et al.*, 2002; Meintjes *et al.*, 2009, 2012):

1. *Sperm capacitation:* sterols are removed from the plasma membrane, and this requires a sterol acceptor molecule (such as albumin) in the medium.
2. *Handling:* a molecule with surfactant properties facilitates sperm and embryo handling by preventing them from sticking to pipettes, dishes, or tubes.
3. Proteins act as lipid and peptide carriers, chelators, cell surface protectors, and regulators of redox potential, and after hydrolysis are a source of fixed nitrogen/amino acids.

Recombinant human serum albumin (HSA) has been available as an equivalent medium supplement since 2002 (Bungun, 2002). Commercially prepared media are often supplied complete, containing a serum substitute such as HSA. Human serum contains approximately 4.5% albumin, i.e. 45 mg/ml. A supplement of 10% HSA provides an albumin concentration of 4.5%. "Serum Substitute Supplement" (SSS) is a 6% protein solution made up of 84% HSA, 16% alpha- and beta-globulins, and a trace of gamma-globulins. SSS is often mistranslated as "Synthetic Serum Substitute," which is misleading. Similarly, companies that label their media "synthetic" in fact mean that it is chemically defined. Commercial media require CE marking for approved use in European countries such as the UK. However, although media without supplements may be CE marked, the serum supplement may not be. This has led some media companies to withdraw supplies of supplemented media, with a recommendation that recombinant HSA, sometimes at a higher concentration, should be added before the media are used.

Some media companies stipulate that only the protein source provided by that company should be used to supplement the media that they supply. Although providing an adequate protein source at the correct level is crucial, the QA provided by the media company may be invalidated if a different protein source is used.

Mineral oil

An oil overlay used to protect medium droplets has been a routine part of human IVF culture since the 1960s (Brinster, 1963; Purdy, 1982; Elder & Johnson, 2015c), and confers specific advantages in maintaining stability of the system (Elder & Dale, 2010, Ch. 8). The layer of oil slows down evaporation of the medium, and acts as a buffer to sudden fluctuations in osmolality, temperature, pH, and gas exchange. Two types of oil have been used: mineral and silicone (see Table 3.1). Silicone oil, used as a stationary phase in gas chromatography or as an anti-foaming agent, should not be confused with mineral oil,

Table 3.1 Types of oil used in culture overlays (adapted from Elder & Dale, 2010, Ch. 8)

1. Paraffin oil – also known as white mineral oil or liquid paraffin
   Merck Index: "Liquid Petrolatum" (Petrolatum = Vaseline)
   - Is derived from petroleum, and exists in many different forms with differing melting points and viscosities. Available in "light" or "heavy" forms. "Light" density = 0.83 to 0.89 g/ml at 25 °C
   - Batch-to-batch variations are common, since they are a mixture of hydrocarbons with fluctuating composition
2. Silicone oil (Dimethyl-polysiloxane) (Erbach *et al.*, 1995)
   - Polymer, available in many viscosities, from 5 to 600 000 centistokes at 25 °C
   - Problems have been encountered with Zn^{2+} contamination, and toxicity after degradation as a result of exposure to sunlight (Provo & Herr, 1998)

a by-product of the petrol industry obtained by fractionated distillation, which contains a large mixture of carbohydrates.

Refined distillates of petroleum that are fractionated into mineral oils are separated into "technical" and "food/medicinal" grades, which undergo more stringent processing; these are used in IVF culture, and have specifications for low levels of polycyclic aromatic hydrocarbons (PAHs), unsaturated hydrocarbons, and aromatic nitrogen and sulfur compounds. Paraffin oil is a type of mineral oil, distinguished from "light" mineral oil by the length of carbohydrate chain present. If the purification process of the oil is not carefully controlled, it can be contaminated with multi-cyclic aromatic, unsaturated, and aromatic hydrocarbons. The unsaturated bonds in polycarbon lipid tails make the oil susceptible to photo-oxidation/peroxidation, with accumulation of peroxide levels that are damaging to embryo culture. In the presence of elevated peroxide levels, albumin in culture medium can facilitate the transfer of free radicals across the zona pellucida, possibly due to capture of free radicals by free sulfhydryl groups in the albumin molecule (Otsuki *et al.*, 2009). Washing oil with medium containing albumin may lower levels of toxic contaminants; albumin also binds heavy metals and removes them from the oil.

Oil is vulnerable to peroxidation during transport or storage, with generation of peroxide levels that are toxic, especially if the oil has been exposed to sunlight or UV light (Provo & Herr, 1998). As part of a troubleshooting exercise to determine possible causes of unexpected embryo deterioration, Otsuki *et al.* (2007) carried out a series of experiments to assess potential toxicity of mineral oil supplies, including peroxide values (POV) during different conditions of storage. Sunlight and UV light caused a dramatic increase in POV (up to 3 meq/kg), and deliberate exposure to raised temperature (equivalent to the inside of a car during summer) in the dark, or to atmospheric oxygen (after opening), also resulted in increased POV (0.1 and 0.21 meq/kg, respectively). Significant peroxide elevation was also detected in unopened mineral oils that had not yet reached their expiration date, with different batches showing different time-dependent increases. The authors suggested that purity of the mineral oils, as well as conditions during storage and transport, could have an effect on the rate at which a batch of oil might reach POV levels that are toxic. They concluded that bottles of oil should be stored at 4 °C, in the dark. Peroxide levels in oils vary among manufacturers and also differ according to storage conditions: oil quality should be monitored carefully and routinely, regardless of supplier.

Bear in mind that, even when packaged and sealed, oil that has passed an MEA and the most stringent control checks can deteriorate during transit and/or storage if exposed to environmental chemicals. Oil that is stored in the vicinity of other chemicals, either at the media company, during transit, or in the clinic, can acquire toxicity, despite the safety data provided on the Certificate of Analysis. Any concerns about toxicity of oil (or any media) should be reported to the media company, in order to establish whether any other clinics have reported similar toxicity for a specific batch. Full transparency from the companies is essential to help uncover the root cause of any oil-related toxicity. Washing oil can reduce levels of toxicity; Morbeck *et al.* (2010) used oil with identified peroxide, aldehydes, and alkenals to compare different washing procedures: media alone, media +5 mg/ml HSA, and distilled water. All were equally effective in reducing toxin levels: the oil containing peroxide passed the one-cell mouse MEA after washing.

A further experiment compared bovine embryo development in cultures overlaid with either washed light mineral oil (Sigma-Aldrich) or sterile light paraffin oil (SPO, Ovoil, Vitrolife) (Tae *et al.*, 2006). Significantly higher development to the morula and blastocyst stage was observed with the use of SPO, and the authors confirmed that oil quality was an important factor in achieving optimal *in vitro* embryo development.

Chemical fingerprinting

Advanced technology that combines Fourier transform infrared spectroscopy (FTIR), nuclear magnetic resonance spectroscopy (NMR), gas chromatography coupled to mass spectrometry (GC-MS), and kinematic viscosity analysis can now be used to examine oil composition in detail. The combined technology gives a comprehensive physicochemical analysis that screens for the presence of potentially embryotoxic substances, as well as unsaturated hydrocarbon chains that might act as targets for free radical attack. Although not currently in routine use by all commercial suppliers of oil for IVF, this "fingerprint" could be used as a means of controlling batch-to-batch variation in samples of mineral oil.

Fig. 3.3 Laboratory desiccator adapted to maintain gas atmosphere for embryo culture; the vessel is flushed with gas mixture and placed in an incubator held at 37 °C (this photograph is of the desiccator that was used in November 1977, to culture the embryo that led to the birth of Louise Brown in July 1978). *Bourn Hall Clinic Archives*

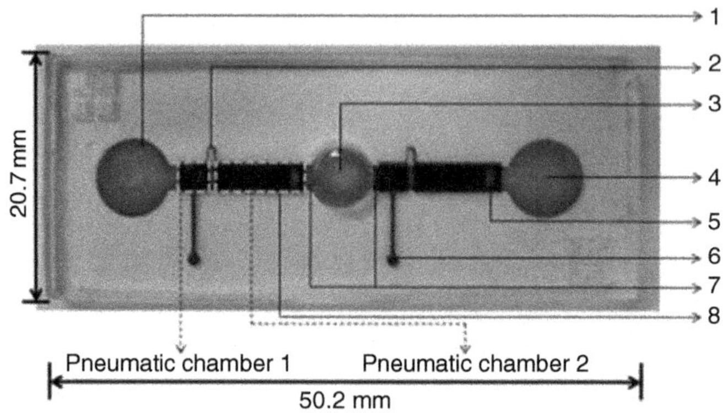

Fig. 3.4 Example of a microfluidic device used for embryo culture.

1. Reservoir for fresh medium
2. Pneumatic microchannel
3. Cell culture chamber
4. Waste medium reservoir
5. Normally closed valve
6. Airtube insertion hole
7. Microchannel for medium
8. Pneumatically driven membrane-based micropump

(Adapted from: Monteiro da Rocha A & Smith GD, 2012, Springer e-book; with kind permission from Springer Science +Business Media.)

New directions

The quest for the Holy Grail of defining better media and culture systems has travelled from simple salt solutions in gas-filled desiccators (Fig. 3.3) to computer-controlled microchambers with continuous renewal of media (Fig. 3.4). Microfluidic systems extend the strategy behind sequential media by aiming to mimic dynamic changes in the tubal environment of the embryo during its transit towards the uterus (Suh *et al.*, 2003; Wheeler *et al.*, 2007; Smith *et al.*, 2011; Swain *et al.*, 2012, 2013; Monteiro da Rocha & Smith, 2012, 2014). At present, the use of such chambers is limited by high costs, and by problems associated with maintaining sterile conditions in pumps and tubing. Dynamic culture platforms that involve tilting and vibrating ("rock and roll") have also been proposed as being of potential benefit, by disrupting media gradients that form around cells or stimulating mechano-sensitive signaling pathways that might stimulate embryo growth (Swain, 2013).

Conclusion

This quest for defining better media and culture systems has become an ongoing process of evolution in synchrony with applying a better understanding of the complex mechanisms that control early human embryo development. It may become clear that the use of a single culture medium or system for gamete culture is partially in discordance with early "individual" embryo requirements, in relation to their wide variation in quality. As systems become more effective at "rescuing" sub-optimal embryos, monitoring the health and development of babies born becomes ever more critical.

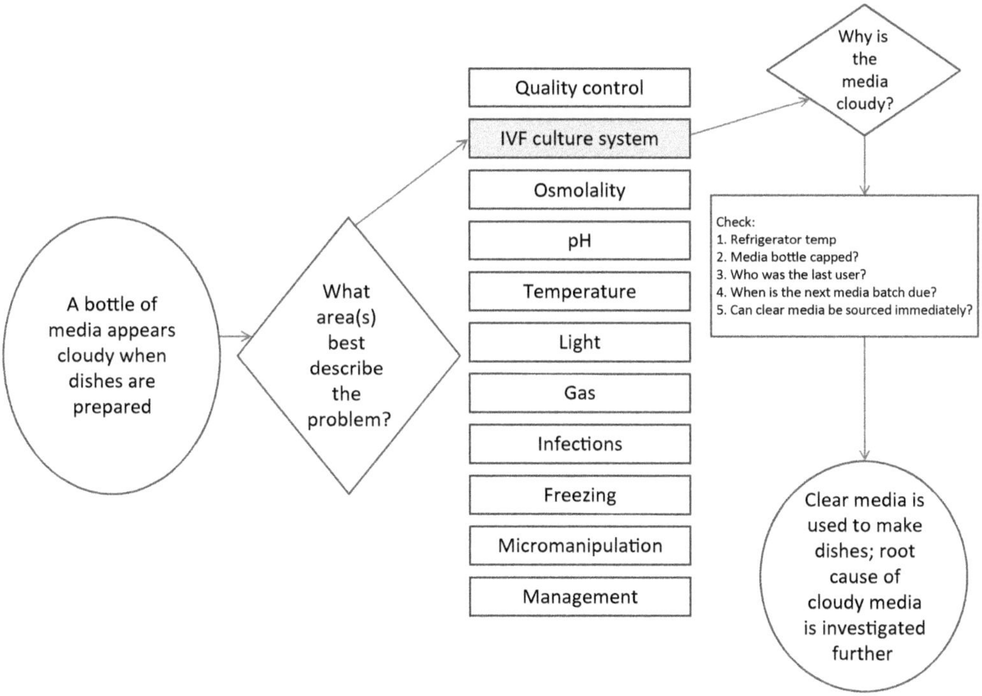

Fig. 3.5 Process map for troubleshooting an IVF culture system.

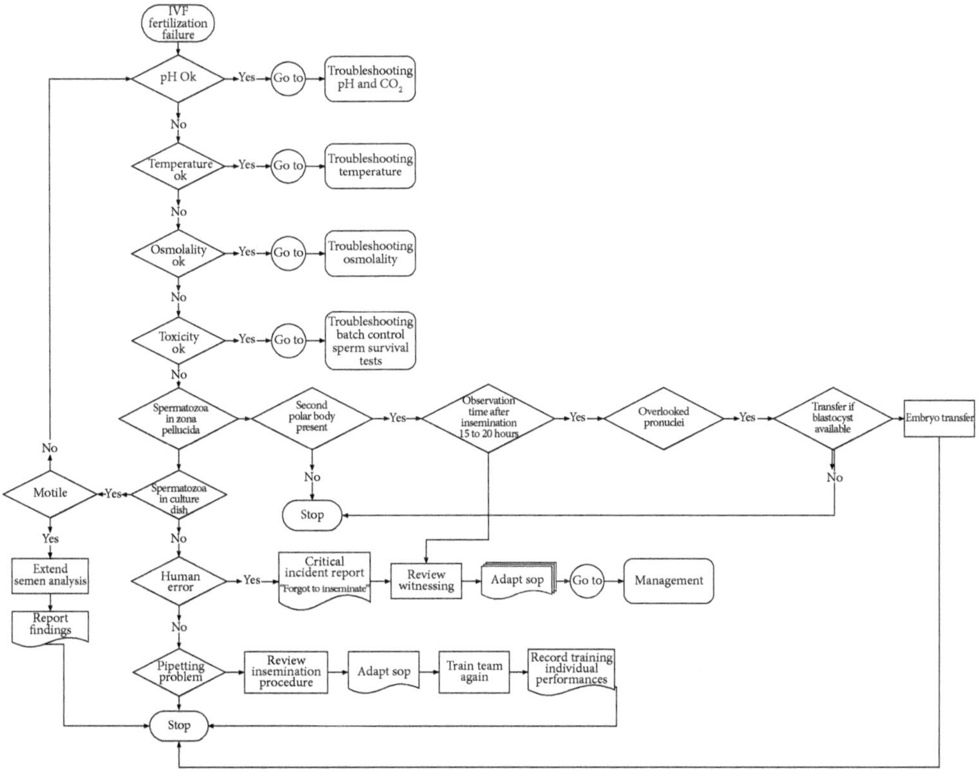

Fig. 3.6 Flowchart for troubleshooting an IVF culture system.

Further reading

Biggers JD (2002). Thoughts on embryo culture conditions. *RBM Online* **4**: 30–8.

P Quinn (ed.) (2014). *Culture Media, Solutions, and Systems in Human ART*, Cambridge University Press.

GD Smith, JE Swain, TB Pool (eds.) (2012). *Embryo Culture: Methods and Protocols* Springer, Berlin.

References

Ali J (2000). Formulation of protein free medium for human assisted reproduction. *Human Reprod*, **15**(1), 145–56.

Ali J (2004). Generation of viable human embryos in a protein-free embryo culture medium enhances clinical pregnancy rate and prevents disease transmission in assisted reproduction. *Middle East Fertil Soc J*, **9**(2), 118.

Anckaert E, Adrianssen T, Romero S, Smitz J (2009). Ammonium accumulation and use of mineral oil overlay do not alter imprinting establishment at three key imprinted genes in mouse oocytes grown and matured in a long-term follicle culture. *Biol Reprod*, **81**(4), 666–73.

Bavister B (1993). *Preimplantation Embryo Development*. Serono Symposia US, Sprenger Verlag, p. 69.

Biggers JD, Rinaldi LM, Webb M (1957). The study of growth factors in tissue culture. *Symp Soc Exp Biol* **11**, 264–97.

Biggers JD (1998). Reflections on the culture of the preimplantation embryo. *Int J Dev Biol*, **42**, 879–84.

Biggers JD (2004a). Discrepancies between the effects of glutamine in cultures of preimplantation mouse embryos. *RBMOnline*, **9**(1), 70–3.

Biggers JD (2004b). Enhanced effect of glycyl-L-glutamine on mouse preimplantation embryos in vitro. *RBMOnline*, **9**(1), 59–69.

Biggers JD, Racowsky C (2002). The development of fertilized human ova to the blastocyst stage in KSOMAA medium: is a two-step protocol necessary? *Reprod Biomed Online*, **5**(2), 133–40.

Biggers JD, Summers MC (2008). Choosing a culture medium: making informed choices. *Fertil Steril*, **90**(3), 473–83.

Blake D, Svalander P, Jin M, Silcersand C, Hamberger L (2002). Protein supplementation of human IVF culture media. *JARG*, **19**(3), 137–43.

Brinster RL (1963). A method for *in vitro* cultivation of mouse ova from the two cell to blastocyst. *Exp Cell Res*, **32**, 205–7.

Bungun M (2002). Recombinant human albumin as protein source in culture media used for IVF: a prospective randomized study. *RBMO Online*, **4**(3), 233.

Campbell JM, Mitchell M, Nottle MB, Lane M (2011). Development of a mouse model for studying the effect of embryo culture on embryonic stem cell derivation. *Stem Cells Dev*, **20**(9), 1577–86.

Chang MC (1967). Effects of progesterone and related compounds on fertilization, transportation and development of rabbit eggs. *Endocrinology*, **81**(6), 1251–60.

Cohen G (1987). The Fenton reaction. In *Handbook of Methods for Oxygen Radical Research*, ed. RA Greenwald. CRC Press, Boca Raton, Florida, pp. 55–64.

Cumming RC, Andon NL, Haynes PA *et al.* (2004). Protein disulfide bond formation in the cytoplasm during oxidative stress. *J Biol Chem*, **279**, 21749–58.

Dale B, DeFelice LJ (1990). Soluble sperm factors, electrical events and egg activation. In *Mechanism of Fertilization: Plants to Humans*, ed. B Dale, Springer, Berlin (NATO ASI Cell Biology series H45).

Dale B, Gualtiery T, Talevi R, Tosti E, Santella L, Elder K (1991). Intercellular communication in the early human embryo. *Mol Reprod Dev*, **29**(1), 22–8.

Dolinoy DC, Huang D, Jirtle RL (2007). Maternal nutrient supplementation counteracts bisphenol A-induced DNA hypomethylation in early development. *PNAS*, **104**(32), 13056–61.

Ebner T, Moser M, Sommergruber M, Shebl O, Tews G (2006). Incomplete denudation of oocytes prior to ICSI enhances embryo quality and blastocyst development. *Hum Reprod*, **21**(11), 2972–7.

Elder K, Dale B (2010). The clinical in vitro fertilization laboratory. In *In Vitro Fertilization*, 3rd edn, Cambridge University Press, Chapter 6.

Elder K, Johnson MH (2015c). The Oldham Notebooks: an analysis of the development of IVF 1969–1978. iii. Variations in procedures (c). *Reprod Biomed and Society*, in press.

Erbach GT, Bhatnagar P, Baltz JM, Biggers JD (1995). Zinc is a possible toxic contaminant of silicone oil in microdrop cultures of preimplantation mouse embryos. *Hum Reprod*, **10**, 3248–54.

Grzelak A, Rychlik B, Bartosz G (2001). Light-dependent generation of reactive oxygen species in cell culture media. *Free Radical Biol Med*, **30**, 1418–25.

Guérin P, El Mouatassim S, Ménézo Y (2001). Oxidative stress and protection against reactive oxygen species in the pre-implantation embryo and its surroundings. *Hum Reprod Update*, **7**, 175–89.

Hammamah S, Ménézo Y (Ed). (1999). Ellipses. In *Ovocyte et embryon: De la physiologie à la pathologie*, Paris.

Hiura H, Okae H, Miyauchi N, *et al.* (2012). Characterization of DNA methylation errors in patients with imprinting disorders conceived by assisted reproduction technologies. *Hum Reprod*, **27**, 2541–8.

Ho Y, Doherty AS, Schultz RM (1994). Mouse preimplantation embryo development *in vitro*: effect of sodium concentration in culture media on RNA synthesis and accumulation and gene expression. *Mol Reprod Dev*, **38**, 131–41.

Hughes PM, Morbeck DE, Hudson SBA, Fredrickson JR, Walker DL, Coddington CC (2010). Peroxides in mineral oil used for in vitro fertilization: defining limits of standard quality control assays. *J Assist Reprod Genet*, **27**(2–3), 87–92.

Jung T (1989). Protein synthesis and degradation in non-cultured and *in vitro* cultured rabbit blastocysts. *J Reprod Fertil*, **86**, 507–12.

Khosla S, Dean W, Reik W, Feil R (2001). Culture of preimplantation embryos and its long-term effects on gene expression and phenotype. *Hum Reprod Update*, **7**, 419–27.

Lane M, Gardner DK (1994). Increase in postimplantation development of cultured mouse embryos by amino acids and induction of fetal retardation and exencephaly by ammonium ions. *J Reprod Fertil*, **102**, 305–12.

Lane M (2003). Ammonium induces aberrant blastocyst differentiation, metabolism, pH regulation, gene expression and subsequently alters fetal development in the mouse. *Biol Reprod*, **69**, 1109–17.

Lane M, Gardner D (2007). Embryo culture: which is the best? *Best Pract Res Clin Obst Gynaecol*, **21**(1), 83–100.

Lawitts JA, Biggers JD (1991). Optimization of mouse embryo culture media using simplex methods. *J Reprod Fertil*, **91**, 543–56.

Leese H (1987). Analysis of embryos by non-invasive methods. *Hum Reprod*, **2**, 37.

Martín-Romero FJ, Miguel-Lasobras EM, Domínguez-Arroyo JA, González-Carrera E, Alvarez IS (2008). Contribution of culture media to oxidative stress and its effect on human oocytes. *RBM Online*, **17**(5), 652–61.

Ménézo Y (1976). Milieu synthetíque pour las survie et la maturation des gamétes et pour la culture de l'oef fécondé. *CR Acad Sic Paris, Ser D*, **282**, 1967–70.

Ménézo Y, Lichtblau I, Elder K (2013). New insights into human preimplantation metabolism *in vivo* and *in vitro*. *JARG* **30**(3), 293–303.

Ménézo Y, Guerin P (2007). Preimplantation embryo metabolism and embryo interaction with the in vitro environment. In *Human Preimplantation Embryo Selection*, K Elder, J Cohen, (eds), Ch. 16, pp. 191–200.

Meintjes M, Chatilis SJ, Ward DC, *et al.* (2009). A randomised controlled study of human serum albumin and serum substitute supplements as protein supplements for IVF culture and the effect on live birth rates. *Hum Reprod*, **24**(4), 782–9.

Meintjes M (2012). Media composition: macromolecules and embryo growth. *Methods Mol Biol*, **912**, 107–27.

Michelson AM (2000). Photochemical production of oxy radicals. In: *Handbooks of Methods for Oxygen Radical Research*, RA Greenwald CRC Press, Boca Raton, FL, pp. 71–5.

Mok-Lin E, Ehrlich S, Williams PL, *et al.* (2010). Urinary bisphenol A concentrations and ovarian response among women undergoing IVF. *Int J Androl*, **33**(2), 385–93.

Monteiro da Rocha A, Smith GD (2012). Culture systems: fluid dynamic embryo culture systems (Microfluidics). In *Embryo Culture: Methods and Protocols, eds* GD Smith, JE Swain, TB Pool (eds). *Methods in Molecular Biology*, **912**, 355–65.

Monteiro da Rocha A, Smith GD (2014). Culture of embryos in dynamic fluid environments. In *Culture Media, Solutions, and Systems in Human ART*, Ed. P Quinn (ed.), Cambridge University Press, Ch. 15, 197–210.

Morbeck D, Khan Z, Barnidge DR, Walker DL (2010). Washing mineral oil reduces contaminants and embryotoxicity. *Fertil Steril*, **94**(7), 2747–52.

Otsuki J, Nagai Y, Chiba K (2007). Peroxidation of mineral oil used in droplet culture is detrimental to fertilization and embryo development. *Fertil Steril*, **88**, 741–3.

Otsuki J, Nagai Y, Chiba K (2009). Damage of embryo development caused by peroxided mineral oil and its association with albumin in culture. *Fertil Steril*, **91**(5), 1745–9.

Pomeroy K, Reed M (2013). The effect of light on embryos and embryo culture. *J Reprod Stem Cell Biol*, **3**(2), 46–54.

Provo MC, Herr C (1998). Washed paraffin oil becomes toxic to mouse embryos upon exposure to sunlight. *Theriogenology*, **49**(1), 214.

Purdy JM (1982). Methods for fertilization and embryo culture in vitro. In *Human Conception* in Vitro; *Proceedings of the first Bown Hall Meeting*, R Edwards & JM Purdy (eds.). Academic Press, London, pp. 135–48.

Quinn P, Kerin JF, Warren GM (1985). Improved pregnancy rate in human *in vitro* fertilization with the use of a medium based on the composition of human tubal fluid. *Fertil Steril*, **44**, 493–8.

Quinn P (2012). Culture systems: sequential. *Methods Mol Biol*, **912**, 211–30.

Quinn P (ed.) (2014). *Culture Media, Solutions and Systems in ART*, Cambridge University Press.

Smith GD, Swain JE, Borman CL (2011). Microfluidics for gametes, embryos and embryonic stem cells. *Reprod Med*, **29**(1), 5–14.

Smith GD, Takayama S, Swain J (2012). Rethinking *in vitro* embryo culture: new developments in culture platforms and potential to improve assisted reproductive technologies. *Biol Reprod*, **86**(3), 1–10.

Summers MC (2014). A brief history of the development of the KSOM family of media. *Hum Fertil* (Camb), **17** Suppl 1, 12–16.

Swain JE, Lai D, Takayama S, Smith GD (2013). Thinking big by thinking small: application of microfluidic technology to improve ART. *Lab Chip*, **13**(7), 1213–24.

Swain JE, Smith GD (2011). Advances in embryo culture platforms: novel approaches to improve preimplantation embryo development through modifications of the microenvironment. *Hum Reprod Update*, **17**(4), 541–57.

Swain J (2013). Shake, rattle and roll: bringing a little rock to the IVF laboratory to improve embryo development. *JARG*, **89**(4), 105.

Suh RS, Phadke N, Ohl DA, Takayama S, Smith GD (2003). Rethinking gamete/embryo isolation and culture with microfluidics. *Hum Reprod Update*, **9**(5), 451–61.

Summers MC (2013). A brief history of the development of the KSOM family of media. *JARG*, **30**(8), 995–9.

Tritsch G (1962). Spontaneous decomposition of glutamine in cell culture media. *Exp Cell Res*, **28**(2), 360.

van Os HC, Drogendijk AC, Fetter WP, Heijtink RA, Zeilmaker GH (1991). The influence of contamination of culture medium with hepatitis B virus on the outcome of *in vitro* fertilization pregnancies. *AJOG*, **165**(1), 152.

Velker BA, Denomme MM, Mann MR (2012). Embryo culture and epigenetics. *Methods Mol Biol*, **912**, 399–421.

Vergouw CG, Kostelijk EH, Doejaaren E, Hompes PG, Lambalk CB, Schats R (2012). The influence of the type of embryo culture medium on neonatal birthweight after single embryo transfer in IVF. *Hum Reprod*, **27**(9), 2619–26.

Wales G, Whittingham D (1967). A comparison of the uptake and utilisation of lactate and pyruvate by one- and two-cell stage mouse embryos. *Biochem Biophys Acta*, **148**, 703.

Wales G, Whittingham D (1973). The metabolism of specifically labelled lactate and pyruvate by two-cell mouse embryos. *J Reprod Fertil*, **33**, 207.

Wheeler MB, Walters EM, Beebe DJ (2007). Toward culture of single gametes: the development of microfluidic platforms for assisted reproduction. *Theriogenology*, **1**, 68 Suppl 1:S178–89.

Whitten WK, Biggers JD (1968). Complete development *in vitro* of the pre-implantation stages of the mouse in a simple chemically defined medium. *J Reprod Fertil*, **17**, 399–401.

Whitten WK (1956). Culture of tubal ova. *Nature*, **177**, 96.

Wrenzycki C, Herrmann D, Carnwath JW Niemann H (1999). Alterations in the relative abundance of gene transcripts in preimplantation bovine embryos cultured in medium supplemented with either serum or PVA. *Mol Reprod Dev*, **53**, 8–18.

Wrenzycki C, Herrmann D, Keskintepe L, *et al.* (2001). Effects of culture system and protein supplementation on mRNA expression in pre-implantation bovine embryos. *Hum Reprod*, **16**, 893–901.

Young L (1998). Large offspring syndrome in cattle and sheep. *Rev Reprod*, **3**, 155.

Zander-Fox DL, Mitchell M, Thompson JG, Lane M (2010). Alterations in mouse embryo intracellular pH by DMO during culture impair implantation and fetal growth. *RBM Online*, **21**(2), 219–22.

Zander-Fox D, Lane M (2012). The future of human embryo culture media – or have we reached the ceiling? In *The Human Embryo*, Yamada S (ed.), InTech, pp. 73–98.

Chapter 4

Osmometry

Introduction

Osmolarity and *osmolality* both measure osmotic pressure, which depends upon the concentration of solutes in a solution; the two terms refer to a different unit of measurement: osmolarity is measured in mOsmoles/**l** (volume) whilst osmolality is measured in mOsmoles/**kg** (weight). Osmolarity measures the *concentration of an osmotic solution*, and osmolality measures the *concentration of particles in a fluid.* In practice, the absolute values of the two measurements differ very little, and for this reason the terms are often used interchangeably. However, it is important to understand the difference between them.

An osmole is one mole (i.e. gram-molecular weight) of any non-dissociable substance. It will contain 6.02×10^{23} (Avogadro's number) osmotically active particles.

1. Osmol**arity** = osmoles of solute particles per litre of solution (mOSm/l).

   Osmotic pressure depends on the total number of solute particles: salts that are completely dissociated in aqueous solution contribute two independent, osmotically active particles, e.g.

NaCl dissociates in solution to one Na^+ ion and one Cl^- ion, so that

1 mol/l of NaCl = 2 Osmol/l

The volume of solution changes with the amount of solute added, as well as with changes in temperature and pressure, and thus osmolarity is difficult to measure.

2. Osmola**lity** = the osmotic concentration of a solution, mOsm/kg of solvent: a 1M solution that contains a single solute dissolved in water will have an osmolality of 1 Osm/kg. The amount (weight) of solvent remains constant regardless of changes in temperature and pressure, and therefore osmolality is more commonly used in practical osmometry, because it is easier to evaluate.

For a given solution, *osmolarity* is slightly less than *osmolality*, because:

- total solution *volume* (used for *osmolarity*) includes solute content
- total solution *weight* (used for *osmolality*) excludes solute weights.

This distinction means that, in osmotic terms, one liter of plasma is not equivalent to one kilogram of plasma: plasma osmolarity and plasma osmolality are different parameters. However, at low concentrations (below approx. 500 mM, i.e. culture media solutions), the weight of the solute is negligible compared to the weight of the solvent, and osmolarity and osmolality are very similar. Osmometers measure "osmolality", expressed as mOsm/kg, and this is the technically correct term that should be used in clinical laboratory practice. *In vitro* culture systems are highly susceptible to changes in both osmolality and osmolarity, both of which are affected by changes in temperature, water content, and pressure.

Culture media osmolality should be similar to that of physiological extracellular fluid (plasma). All cells, including gametes and embryos, are surrounded by cell membranes that are largely permeable to water, but impermeable to most of the solutes present in the cytoplasm or extracellular fluid. If the total concentrations of osmotically active solutes (osmolytes) differ between intracellular and extracellular compartments, there will be a net force on the membrane; because the membranes are pliable, the cell will swell or shrink until intracellular and extracellular osmotic pressures are equal. As a result, through diffusion and osmosis, the composition of extracellular fluid will affect the ionic concentration (osmolality) of the cell cytoplasm.

Diffusion: molecules mix due to the kinetic energy of random motion, going down a concentration gradient until concentrations are equal.

Osmosis: internal energy of solvent molecules drives them to create a more selective diffusion against a concentration gradient.

Osmotic pressure: the hydrostatic pressure (available energy per unit volume) that is caused by a difference in the amounts of solutes between solutions that are separated by a semi-permeable membrane, e.g. the cell membrane of gametes and embryos.

Culture media osmolality

In vitro, gametes and embryos are very sensitive to even small external environmental increases in osmolality, with potentially dramatic effects on viability: inappropriate handling of the culture system can lead to increases that are sufficient to jeopardize embryonic development. Intracellular osmotic pressure regulates cell volume, a crucial factor involved in cellular homeostasis and cell division; early embryos will arrest if their ability to control intracellular volume is disrupted by suboptimal culture conditions. When water evaporates from droplets, the osmolality rises; for this reason, in situations where an oil overlay is not used (e.g. freezing/vitrification), small droplets must be plated immediately before use in order to avoid evaporation, which will increase osmolality. Evaporation is less pronounced when larger volumes (1–2 ml) are used. Where an oil overlay is used (e.g. for culture media), the depth of the oil layer should provide a balance between minimizing evaporation and allowing gas penetration to stabilize pH. In droplet culture, the shape of the droplet under oil also has an influence: a convex shape will be more prone to evaporation at the top surface, where the oil overlay is at its minimum, whereas "flatter" droplets with reduced curvature have a more even interface with the oil.

Osmolality is determined by total solute content; all embryo culture media contain the same inorganic ions: Na^+, K^+, Ca^{2+}, Mg^{2+}, Cl^-, SO_4^{2-}, HCO_3^- and most also contain PO_4^{2-}. During the first half of the 20th century, somatic cells were cultured in basic saline solutions such as Krebs–Ringer bicarbonate (KRB). Mouse embryos were successfully cultured in Whitten's medium (Whitten & Biggers, 1968), which has a similar composition to KRB, with the addition of glucose and bovine serum albumin. M16, a culture medium formulated in the 1970s for mouse embryo culture, is also similar, with Na^+Cl^- replaced by Na^+ lactate. Edwards and Purdy initially cultured human embryos in follicular fluid diluted with Tyrode's solution before trying numerous different combinations of

Table 4.1 Osmolarity of medium originally used for mouse embryo culture

Medium	mOsm/kg	Author
M16 (= Modified Brinster)	290	Whittingham (1971)
CZB	275	Chatot *et al.* (1989)
KSOM	250	Lawitt & Biggers (1991)

somatic cell media (Elder & Johnson, 2015c). Embryos cultured in Earle's Balanced Salt Solution (EBSS) and Ham's F-10 medium, supplemented with serum or serum albumin, led to the first successful implantations. Media formulations for human embryo culture continued to be developed during the 1980s, with the important observation that embryos suffered developmental blocks in media with osmolalities higher than 300 mOsm/kg. Significant changes in osmolality were thought to contribute to developmental arrest of mouse embryos at the two-cell stage (Wang *et al.*, 2011), bovine embryos at the eight-cell stage (Camous *et al.*, 1984), and human embryos at the four- to eight-cell stage (Bolton *et al.*, 1989).

Osmolality was lowered primarily by reducing the concentration of Na^+Cl^- and other inorganic ions, and these experiments with media formulations suggested that optimal development requires a range of 255–295 mOsm/kg (see Table 4.1). Lowering total inorganic ion concentration and thus osmolality was confirmed as a key feature that eliminated *in vitro* blocks to embryo development (Chatot *et al.*, 1989). CZB (Chatot–Ziomek–Bavister) medium supported mouse embryo development at an osmolality of 275 mOsm/kg (Chatot *et al.*, 1989), and blastocyst development was demonstrated at the lower end of the osmolality range (250 mOSm/kg) in KSOM (Potassium Simplex Optimized Medium) (Baltz & Tartia, 2010; Lawitts & Biggers, 1991a, 1991b). Continued development of human embryos to the blastocyst stage requires an *in vitro* osmolality of <300 mOsm/kg (Baltz, 2014).

The addition of amino acids to culture media provided a subsequent improvement to both mouse and human embryo culture, due to their nature as organic osmolytes. Organic osmolytes are small neutral organic compounds that do not disrupt cellular physiology or biochemistry even at high concentrations. The beneficial effect of amino acids during cleavage stages of development may be partly due to mechanisms for cell volume regulation that depend on their accumulation as organic osmolytes to provide intracellular osmotic support. These amino acids, principally glycine and taurine, replace a portion of the intracellular inorganic ions that would otherwise be needed to maintain cell size, preventing intracellular ionic strength from rising to deleterious levels that might block development (Dawson *et al.*, 1998; Dumoulin *et al.*, 1997).

The optimum salt levels, osmolality, and amino acid content of culture media are not independent, but interact strongly because of their roles in cell volume regulation. In the absence of compounds that can be used as organic osmolytes, preimplantation embryos will develop only at lower medium osmolalities and salt concentrations. However, when organic osmolytes are present (e.g. some amino acids), embryos will develop in culture at higher osmolalities, similar to those found in tubal fluid *in vivo*. The average osmolality of tubal and uterine secretions is now thought to range from 275–305 mOsm/kg. In rabbit, sheep, and cow reproductive tracts, tubal secretions have osmolalities of 290 mOsm/kg and

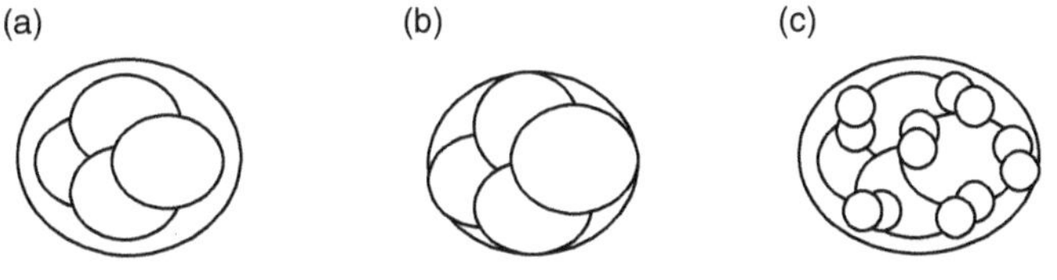

Fig. 4.1 Response of a four-cell embryo to an external hypotonic solution. Fragment extrusion reduces the cytoplasmic volume of individual blastomeres, but vital cytoplasmic components such as mRNA and proteins may be lost into the fragments. (a) Blastomere of normal size in medium with physiological osmolality. (b) When placed in medium with lower osmolality, blastomeres take in water to equalize the osmotic pressure, and initially swell. This may be accompanied by slight expansion of the zona. (c) In order to correct the abnormal increase in volume, the embryo extrudes cytoplasmic fragments. The cytoplasm may then reduce to its original volume, but vital cytoplasmic components such as mRNA and proteins may be lost with the fragments.

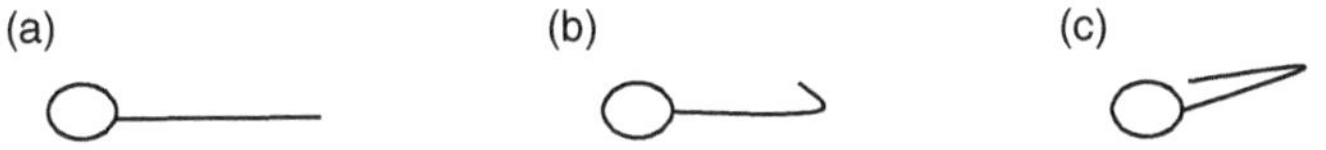

Fig. 4.2 The hypo-osmotic swelling test, used to ascertain whether a sperm is alive or dead (vitality). Immotile sperm are transferred to a hypo-osmotic solution; after approximately 5 minutes (often sooner), the sperm tails begin to swell, as medium is taken up across the tail membranes. Swelling causes the tail to curl, and the degree of curling is related to vitality: the most viable sperm cells exhibit full tail curling. (a) Sperm cell with no tail curling: non-viable. (b) Sperm cell with partial tail curling: viable. (c) Sperm cell with full tail curling: fully viable.

osmolalities of uterine secretions in these species reach up to 315 mOsm/kg, due to a higher concentration of amino acids (Ménézo & Laviolette, 1972; Casslén, 1987; Ménézo & Guerin, 1997).

Intracellular volume can be controlled via mechanisms that actively adjust the cytoplasmic concentrations of osmotically active solutes that will alter intracellular osmotic pressure. During cell division, the nucleocytoplasmic ratio is extremely important (Edgar & Schubiger, 1986): dilution of specific factors initiates a kinase cascade that triggers mitosis. The nucleocytoplasmic ratio governs formation of the Nuclear Organizing Region (NOR), which induces changes in DNA conformation that allows binding of transcription promoters and enhancers. In conditions of low osmolality, the intracellular volume increases, and blastomeres swell. In order to achieve the nucleocytoplasmic ratio necessary for entry into mitosis, the embryo may therefore attempt to lose cytoplasm by making blebs (fragmentation). This results not only in loss of cytoplasm, but also of mRNA and proteins necessary for further development. Fragmentation may thus be an indication of suboptimal culture conditions, not *a priori* associated with apoptosis. However, fragmentation may eventually lead to apoptosis by loss of intolerable levels of cellular material (Fig. 4.1).

Inorganic salts pass freely into the cell cytoplasm, and in conditions where osmolality is dependent mainly on inorganic salt concentrations, changes in water content have a dramatic effect. When an extracellular compartment (culture media) is hypotonic, the cell will swell and may eventually lyse or burst. This is the basis of the hypo-osmotic swelling (HOS) test used to test membrane integrity of sperm, and thus check their viability (Fig. 4.2).

Extracellular hypo-osmotic pressure also affects the concentration of inorganic ions in the cytoplasm, causing a "salting effect" that denatures enzymes and affects macromolecular assembly. In contrast, very high concentrations of extracellular organic osmolytes have very little effect. Active transport of organic osmolytes into the cell increases intracellular pressure, which restores cell volume after an unwanted decrease. Organic osmolytes that

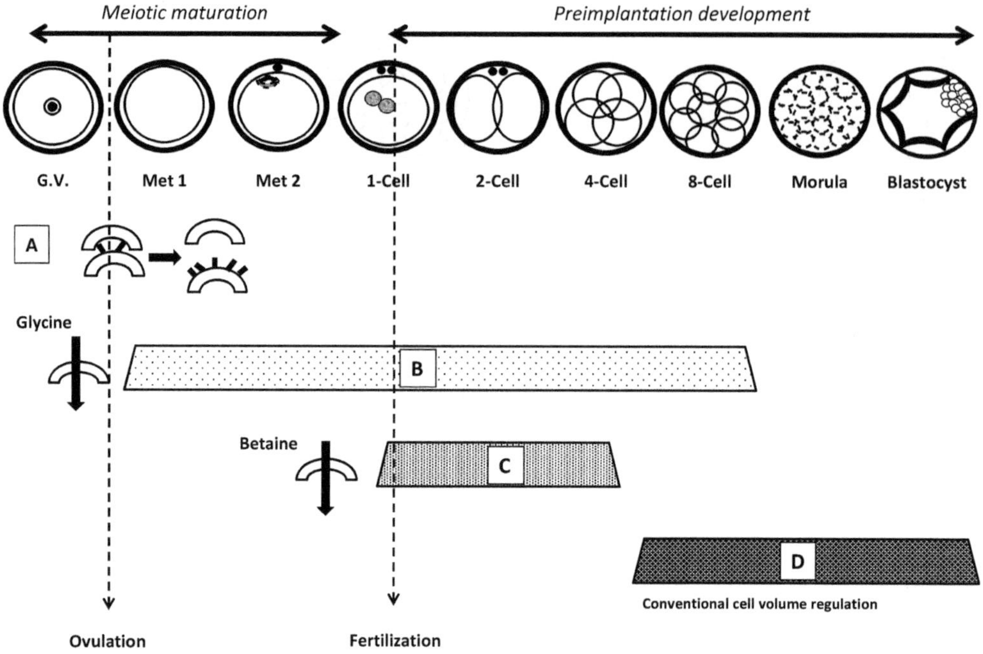

Fig. 4.3 Model of cell volume regulation in oocytes and preimplantation embryos. Developmental timeline is indicated schematically at the top, showing stages of oocyte maturation and preimplantation development; arrows indicate approximate timing of the initiation of ovulation and fertilization. Independent cell volume regulation is initiated just after ovulation, with release of the adhesions between oocyte and zona pellucida (A), followed by activation of glycine transport via GLYT1 (B). The glycine-dependent mechanism controls cell volume, starting during meiotic maturation and ending around the four-to eight-cell stage. A second mechanism of volume regulation via betaine transport and accumulation may function in cell volume homeostasis after fertilization, ending at the two-cell stage (C). These mechanisms of cell volume regulation are unique to oocytes and early embryos; more conventional mechanisms of cell volume regulation may become activated around the time of compaction and morula formation (D).*(Adapted with permission from: Baltz J & Tartia A (2009). Human Reproduction Update, Vol.16, No. 2 pp. 166–176, 2010.)*

are compatible with biochemical function, such as glutamine, taurine, and glycine, provide the intracellular osmotic support required to maintain cell volume without disrupting cell function. Although some organic osmolytes are synthesized within cells (Guérin & Ménézo, 1995), the majority are actively imported via specialized transporters that are activated by increased external osmolality or reduced cell volume, such as the GLYT1 transporter in early mouse embryos. Glycine added to the medium has been shown to protect embryos against increased salt concentrations/osmolality; active transport via GLYT1 is functional only during a brief window, before the onset of compaction. Glutamine, taurine, betaine, proline, and ß-alanine are recognized organic osmolytes, which protect against detrimental effects of increased salt concentration in media.

An additional system that depends on betaine or proline transport may also operate in parallel with the glycine-dependent mechanism for a short period after fertilization. Baltz & Tartia (2009) have proposed a model of cell volume regulation in oocytes and early embryos with a timeline that delineates active transport of glycine and betaine via their transporter molecules. Glycine displaces inorganic ions, and this mechanism allows development at lower osmolality. The presence of glutamine also increases tolerance to higher osmolalities; in the absence of amino acids, embryos will no longer develop (Fig. 4.3).

Due to concerns regarding the detrimental effect of ammonium ions as a metabolic by-product, commercial media preparations generally favor lower osmolality instead of increasing the concentration of some amino acids. Table 4.2 shows the osmolality range of commercial media that is currently available (Baltz, 2012).

Glucose is not the primarily preferred energy source for the early preimplantation embryo: there is metabolic block at the level of glucose-6-phosphate or fructose-6-phosphate isomerase enzymes, and excess glucose uptake can lead to the accumulation of glycogen. This causes water to be absorbed, with an increase in cell volume. In order to maintain an appropriate nucleocytoplasmic ratio, embryos may try to recover by extruding cytoplasmic fragments (see Fig. 4.1), and may also try to reduce the water uptake and maintain homeostasis by exporting taurine via protein kinase C, a Na-taurine co-transporter (Hammamah & Ménézo, 1999). Excess glucose uptake at early stages of development may also result in generation of free radicals; reactive oxygen species can cause serious dysfunction and DNA damage, with blebbing of areas of cytoplasm in attempts to compensate/repair.

Culture dish preparation

Osmolality of a culture system is influenced by the methods used in dish preparation. Minimizing the risk of evaporation is crucial when preparing culture dishes: time, temperature, air flow, volume of media, and oil overlay, as well as the method of preparation, must all be considered in order to avoid osmolality increases that may jeopardize embryo development. Evaporation can be minimized by taking culture media directly from the refrigerator immediately prior to dish preparation, so that the droplets are prepared with cold medium. In open cultures (without oil), a humidified atmosphere in the incubator also prevents evaporation and increase in osmolality.

Swain *et al.* (2012) measured osmolality in a series of experiments that involved different methods of droplet preparation:

1. Standard method, droplets pipetted onto a dish, and then covered with oil.
2. Droplets plated, covered with oil, and then aspirated; original droplets replaced with new droplets of fresh medium beneath the overlay (Fig. 4.4).

The results showed that preparing droplets at 37 °C using either of the two strategies significantly increased osmolality. Osmolality also increased relative to droplet size: the smaller the droplet volume, the higher the risk of increasing osmolality. Preparing very small droplets at 37 °C increased the initial osmolality significantly beyond the level at which embryos can develop (Fig. 4.5).

The contributions of humidity and an oil overlay in maintaining osmolality were demonstrated in experiments carried out using a multi-well dish containing 0.5 ml of medium without oil overlay: after 72 hours incubation at 37 °C, osmolality increased from 280 mOsm/kg to 321 mOsm/kg. Covering the medium with a layer of 0.2 ml oil prevented this increase; adding 3 ml of sterile water to the central reservoir limited the increase in osmolality to 292 mOsm/kg. An additional reservoir of culture medium may be used instead of sterile water. The dish reservoir can be created either in the central space in four-well dishes (in-between the wells) or in the separate outer ring of round cell culture dishes that contain a separate center well. Smaller increases were obtained when gas mixtures were bubbled through 50 ml bottles of medium: osmolality increased above 290 mOsm/kg if this bubbling exceeded 15 minutes (Fig. 4.6).

Table 4.2 Osmolality of commercial media (Adapted from J. Baltz, 2012, with permission)

Company	Medium	Osmolality range (mean) mOsm/kg	Culture stages
LifeGlobal®	Global®	260–270 (265)	All
	HTF	280–292 (286)	Cleavage
	Blastocyst	260–270 (265)	8c to blastocyst
Origio (MediCult)	Universal IVF	277–293 (285)	Cleavage
	ISM1™	272–288 (280)	Cleavage
	ISM2™	272–280 (282)	8c to blastocyst
	EmbryoAssist™	272–280 (282)	Cleavage
	BlastAssist®	272–280 (282)	8c to blastocyst
VitroLife	G-1 (G5 series™)	256–266 (261)	Cleavage
	G-1 with HSA (G5 series™)	249–259 (254)	Cleavage
	G-2™ (G5 series™)	255–265 (260)	8c to blastocyst
	G2™ with HSA (G5 series™)	248–258 (253)	8c to blastocyst
Cook® IVF	Sydney IVF Cleavage	285–295 (290)	Cleavage
	Sydney IVF Blastocyst	285–295 (290)	8c to blastocyst
Cooper Surgical (SAGE®)	Quinn's Advantage® Cleavage	257–273 (265)	Cleavage
	Quinn's Advantage® Blastocyst	257–273 (265)	8c to blastocyst
Fertipro	Ferticult™ IVF	275–285 (280)	Cleavage
	Ferticult™ G3	270–290 (280)	8c to blastocyst
GyneMed	GM501	270–290 (280)	All
InVitroCare	IVC-ONE™, IVC-TWO™	270–290 (280)	Cleavage
	IVC-THREE™	270–290 (280)	8c to blastocyst
Irvine Scientific	SSM™	260–270 (265)	Cleavage
	P1®	282–298 (290)	Cleavage
	ECM®	282–295 (288)	Cleavage
	Multiblast®	281–291 (286)	8c to blastocyst
	HTF	272–288 (280)	Cleavage

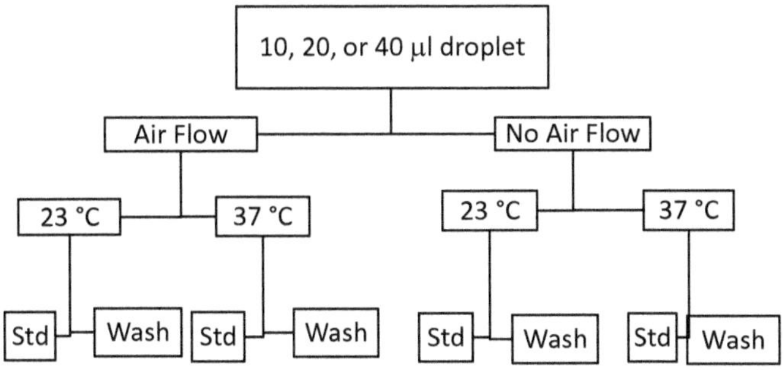

Fig. 4.4 Experimental design: Effect of air flow on osmolality in microdroplet culture. Three different volumes of microdroplets were tested. Drops were made in a sterile hood with or without airflow, with or without a heated surface and using one of two preparation methods. Eight groups of four were tested for each microdroplet volume (24 experimental groups). Media were sampled at 5 minutes and 24 hours, resulting in a total of 48 endpoint measurements. Each measure was performed three times over three replicates and the mean of the nine measurements taken. Std = standard method; Wash = wash-drop method. *(Adapted from: Swain JE, Cabrera L, Smith GD (2012). Reprod BioMed Online,* ***24****(2), 143, with kind permission from Springer Science + Business Media.)*

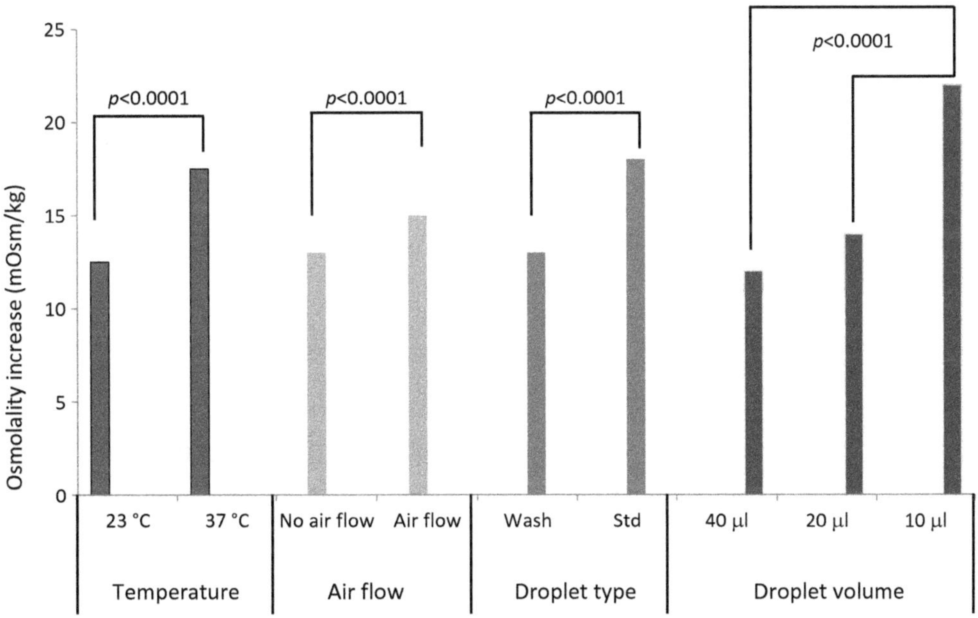

Fig. 4.5 Changes in microdroplet osmolality according to method of preparation.
Conditions used during microdrop preparation have an impact on media osmolality. Data represent sampling after 5 minutes. Lower drop volume, higher temperature, and the standard method of preparation all significantly increase media osmolality. No differences were apparent after 24 hours. Data are mean ± SEM increase over control media sampled out of the bottle. *(Adapted from: Swain JE, Cabrera L, Smith GD (2012). Reprod BioMed Online,* ***24****(2), 143, with kind permission from Springer Science + Business Media).*

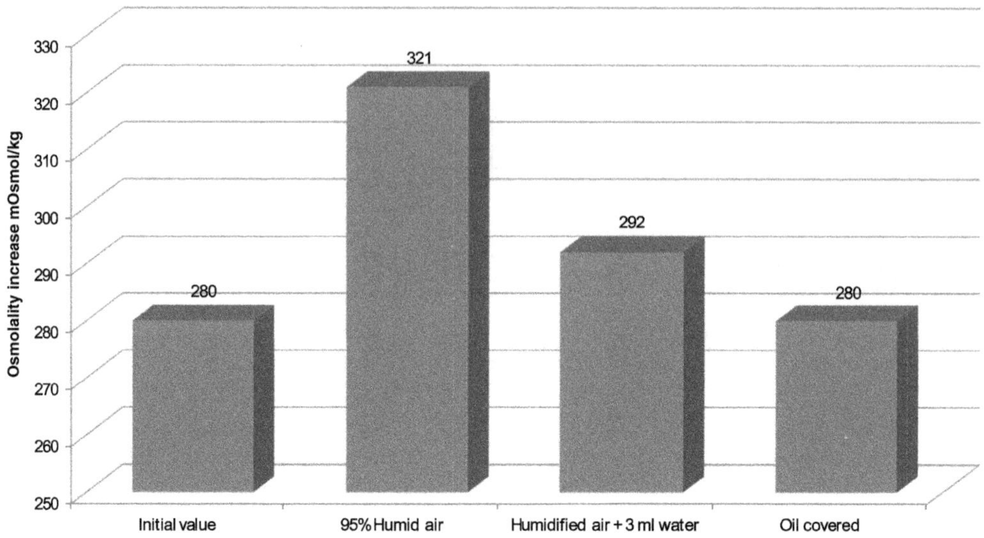

Fig 4.6 Experiment to demonstrate changes in osmolality of 250 μl of culture media in a covered four-well dish incubated at 37 °C for 72 hours. (a) Initial value = 280 mOsm/l. (b) Incubator atmosphere of 95% humidity, Osmolality = 321 mOsm/l. (c) 95% humidity + 3 ml sterile water added to central reservoir of the dish, Osmolality = 292 mOsm/l. (d) Media overlaid with 250 μl oil, Osmolality = 280 mOsm/l. *(Quartec GmbH.)*

A number of systems are now available that differ from the traditional methods of placing droplets onto flat dish surfaces, and osmolality is an important factor that must be taken into consideration when using these dishes. Microwell dishes with indents manufactured into the surface (Well-of-the-Well, Vitrolife, Sweden), GPS Corral dishes (IVFOnline, USA), EmbryoScope dishes (Vitrolife, Sweden), and EEVA dishes (Auxogyn, USA) (Fig. 4.7). The depth of medium, surface area exposed to the oil overlay, and depth of the overlay may all influence the osmolality of the culture medium. Preparation of dishes for time-lapse imaging requires special care in order to make sure that no bubbles are present (see Fig. 4.8). Devices used for microfluidic culture are particularly vulnerable to detrimental osmolality shifts due to evaporation; thin hybrid polydimethylsiloxane (PDMS)-Parylene membranes are recommended to circumvent this problem (Heo *et al.*, 2007).

Practical tips for keeping osmolality constant during preparation of dishes

1. Take note of the osmolality and osmolyte content of the medium to be used; osmolytes act to maintain fluid balance within oocytes/blastomeres, and some media may have a lower concentration of osmolytes than others (HTF has no osmolytes). If there is a risk of increase in osmolality during preparation, consider using a medium that has a lower osmolality.
2. Work at room temperature, using medium that has been taken out of the refrigerator immediately prior to dish preparation.
3. Consider droplet size. Using larger volumes of media in the culture system will minimize the chance of large shifts in osmolality. On this basis, droplet sizes less than 20 μl are not recommended.

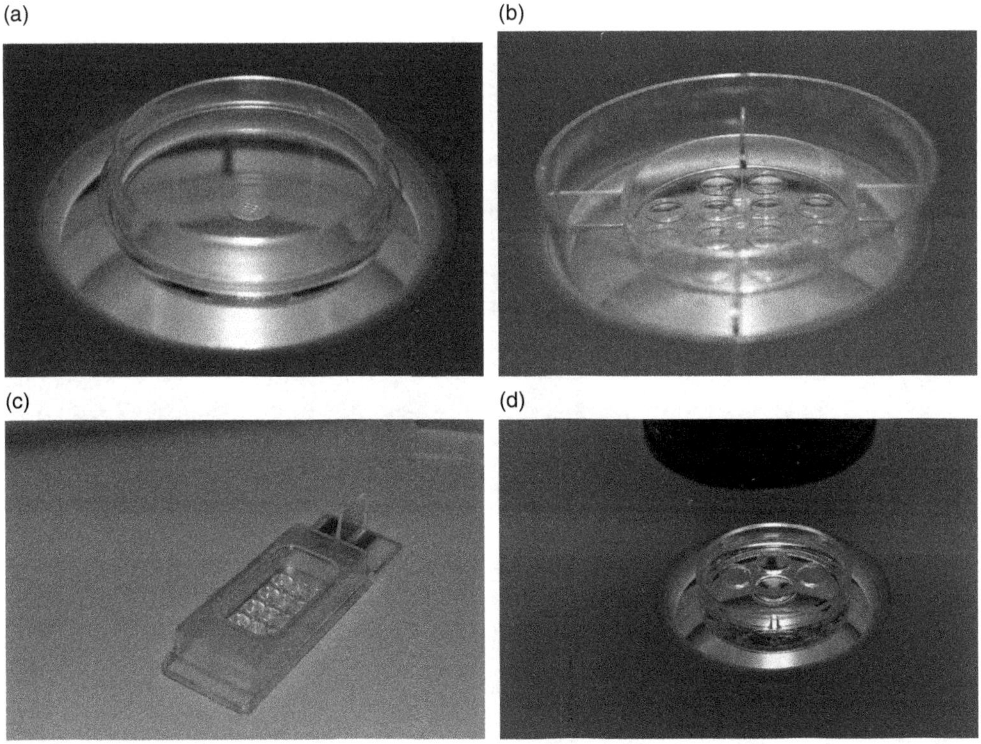

Fig. 4.7 Examples of different types of dishes used for human embryo culture. (a) Well-of-a-well (WOW) dishes contain 9- or 16-microwells, for use with Primo Vision time-lapse systems (Vitrolife); (b) μDrop GPS® Dish contains multiple 20 μl micro-wells, and requires less oil to cover drops than a conventional 60 mm diameter dish (IVFOnline); (c) EmbryoScope dish contains 12 individual microwells for use with EmbryoScope time-lapse systems (Vitrolife); (d) Early embryo viability assessment (EEVA) dish contains 12 microwells for use with EEVA time-lapse systems (Auxogyn). *(Images kindly provided by the Hewitt Fertility Centre, Liverpool Women's Hospital, UK.)*

4. Carefully note the date of first opening the bottle of medium in relation to the date of dish preparation: osmolality can change if the bottle has been opened repeatedly.
5. Note the time/date of dish preparation: the likelihood of evaporation is higher in dishes that have been in the incubator for long periods, resulting in a rise in osmolality.
6. Prepare droplets as quickly as possible: the number of droplets per dish may need to be reduced accordingly.
7. Replace original droplets with droplets of fresh medium in exactly the same spot after overlaying with oil.
8. If sufficient staff are available, the dishes can be prepared by two technicians: one to pipette the droplets into the dish, and the second to immediately cover the droplets with oil. Ensure that good laboratory practice and sterility is rigorously maintained throughout.
9. Be aware that osmolality will change more rapidly with "open" culture, and without humidified incubators.
10. Reduce airflow during dish preparation by turning the laminar air flow off. However, this is recommended only if background air has a suitable grade of purity. Since it is essential that sterility is maintained during dish preparation, an alternative to consider is reducing the air flow velocity by setting it to the lowest level during dish preparation.

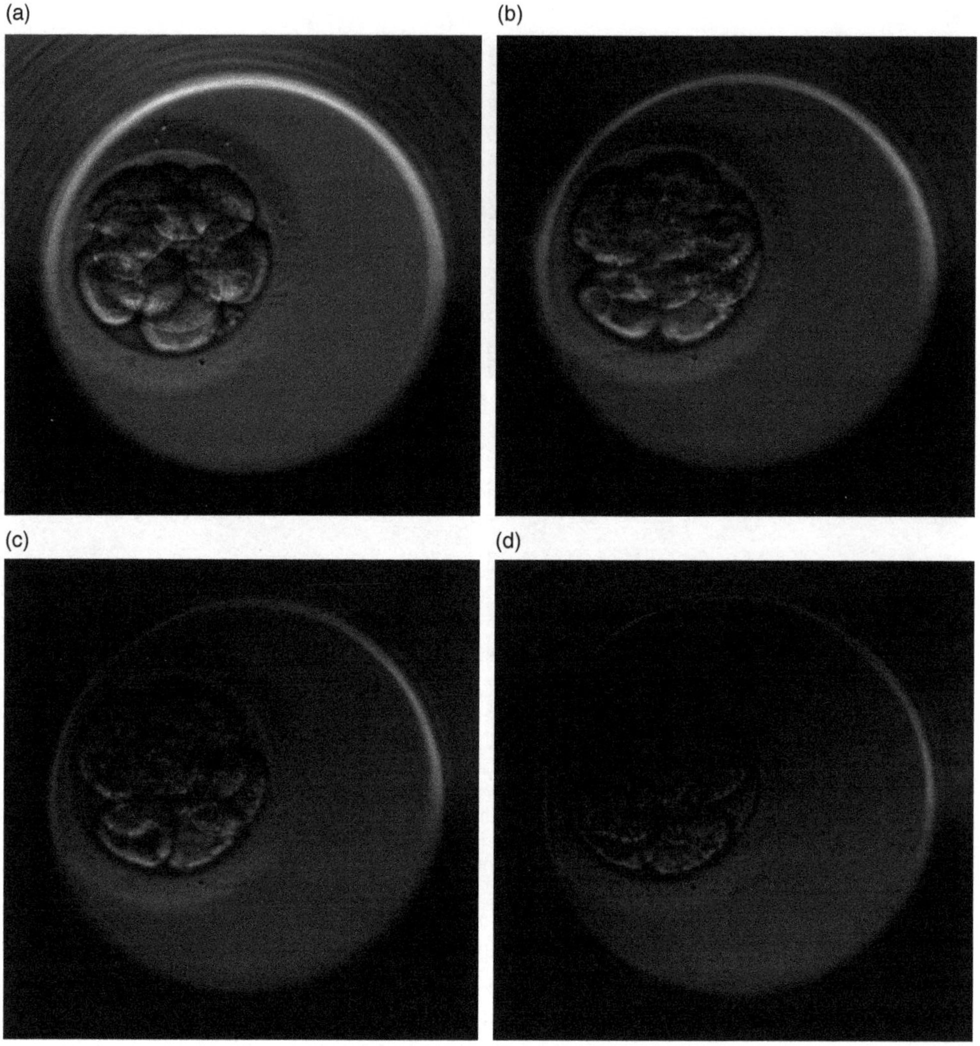

Fig. 4.8 Four images that follow a Day 3 embryo via time-lapse. As progressive images are captured ((a) – (c)), an air bubble moves directly over the top of the microdroplet, finally obscuring the image (d). This can be prevented by carefully inspecting time-lapse dishes and removing any air bubbles seen before placing them in position for imaging. (*Images kindly provided by the Hewitt Fertility Centre, Liverpool Women's Hospital, UK.*)

11. Use washed/humidified oil: in general, there is always an equilibrium between the components of two liquid phases. Unless the oil is first washed with the same medium, an oil interface with albumin-carrying lipids in the medium may extract all lipophilic elements present in the medium. Washing can be carried out by gently shaking oil with medium in a ratio of 1(oil):2(medium) and leaving the mixture to settle overnight, until the two phases are clearly separate.

Commercial media companies usually provide pre-washed oil that has been endotoxin tested and monitored for quality control (QC) via tests such as sperm viability toxicity testing and a mouse embryo assay. Technically, washing the oil after receipt into the

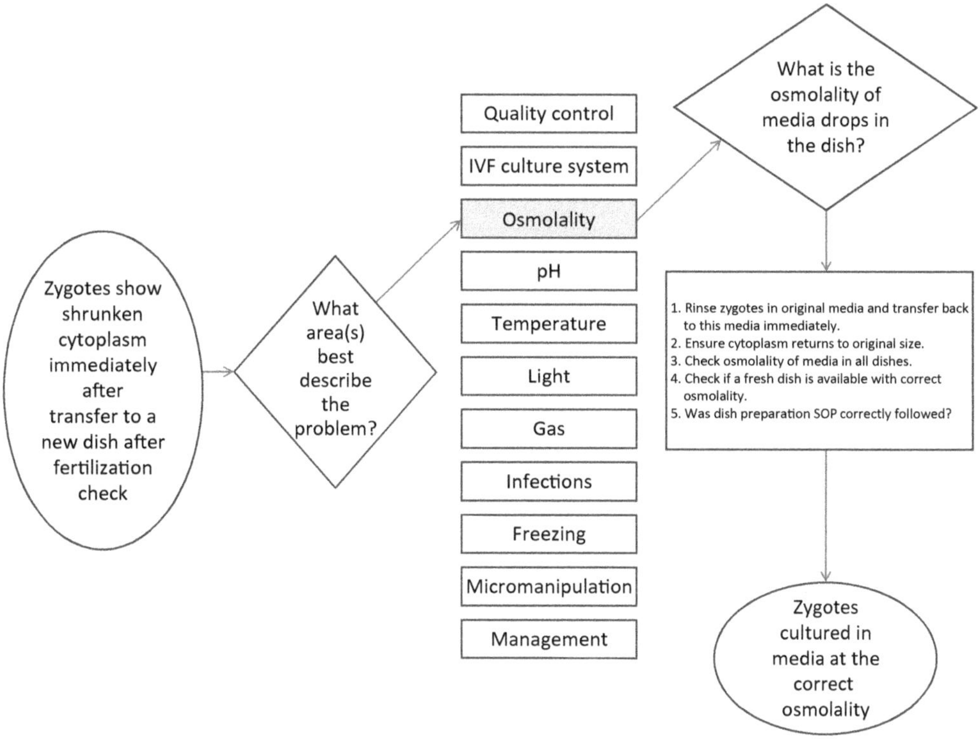

Fig. 4.9 Process map for troubleshooting osmolality problems.

laboratory may invalidate the QC tests, making the clinic responsible for any problems that might arise as a result of additional washing. The company supplying the oil has no control over which medium is used for washing (i.e. whether it is fresh, how long it has been stored, etc.), and thus any warranty originally provided is effectively invalidated. This should be taken into consideration if oil that has been supplied "pre-washed and QC'd" is then washed with media from a different supplier.

Osmolality and cryopreservation

During preparation for cryopreservation, oocytes and embryos are transferred through cryoprotectant solutions with osmolalities that differ from standard culture medium. For slow freezing, the cells are transferred into cryoprotectant solutions of increasing osmolyte concentration, which facilitates cytoplasmic shrinkage. Upon thawing, oocytes and embryos are transferred into solutions of decreasing osmolyte concentration, as the cytoplasmic volume is restored to normal.

Measuring osmolality

Three methods are available to measure osmolality: freezing point depression, vapor pressure depression, and semi-permeable membrane measurements. The use of freezing point depression involves a struggle with inadequate freezing, or undercooling – a common

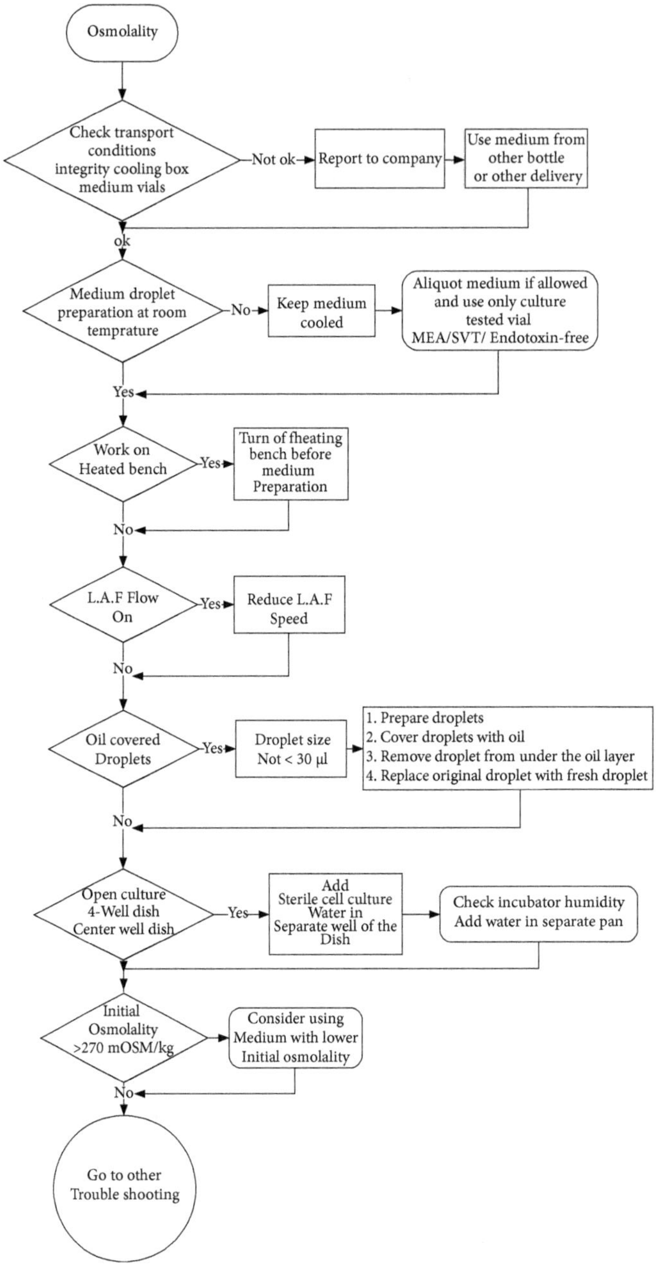

Fig. 4.10 Flowchart for troubleshooting osmolality problems.

phenomenon in very clean and sterile aqueous solutions, and the reason why seeding is used in cryobiology to start ice formation at the proper freezing point (fusion temperature). However, as long as each test is carried out with calibration standards to provide QC, a freezing point osmometer should provide adequate results. Other types of instruments are available, but are seldom used for reasons of costs and availability.

Conclusion

Embryologists should always be aware that osmolality quickly changes during handling of media, with possible detrimental stress to cultured gametes and embryos. An osmolality that is lower than the acceptable range will cause embryos to swell, so that they may temporarily (and deceptively) appear to be of a higher grade. Significant shifts can prevent normal embryo growth and development. Although many laboratories do not routinely use an osmometer, the ability to recognize, identify, and correct any problems that might be related to osmolality is an important factor in troubleshooting, particularly when changes to culture conditions are introduced.

References

Baltz J, Tartia A (2010). Cell volume regulation in oocytes and early embryos: connecting physiology to successful culture media. *Human Reprod Update*, **16**(2), 166–76.

Baltz JM (2012). Media composition: salts and osmolality. *Methods Mol Biol*, **912**, 61.

Baltz JM (2014). Osmolality. In *Culture Media, Solutions and Systems in Human ART*, ed. P Quinn, Cambridge University Press, Chapter 10, pp. 132–41.

Bolton VN, Hawes SM, Taylor CT, Parsons JH (1989). Development of spare human preimplantation embryos *in vitro*: an analysis of the correlations among gross morphology, cleavage rates, and development to the blastocyst. *J In Vitro Fertil Embryo Transfus*, **6**, 30–5.

Camous S, Heyman Y, Meziou W, Menezo Y (1984). Cleavage beyond the block stage and survival after transfer of early bovine embryos cultured with trophoblastic vesicles. *J Reprod Fertil*, **72**, 479–85.

Casslén BG (1987). Free amino acids in human uterine fluid. Possible role of high taurine concentration. *J Reprod Med*, **32**, 181–4.

Chatot CL, Ziomek CA, Bavister BD, Lewis JL, Torres I (1989). An improved culture medium supports development of random-bred 1-cell mouse embryos *in vitro*. *J Reprod Fertil*, **86**, 679–88.

Dawson KM, Collins JL, Baltz JM (1998). Osmolarity-dependent glycine accumulation indicates a role for glycine as an organic osmolyte in early preimplantation mouse embryos. *Biol Reprod*, **59**, 225–32.

Dumoulin JC, van Wissen LC, Menheere PP, Micheils AH, Geraedts JP, Evers JL (1997). Taurine acts as an osmolyte in human and mouse oocytes and embryos. *Biol Reprod*, **56**(3), 739–44.

Edgar BA, Schubiger G (1986). Parameters controlling transcriptional activation during early *Drosophila* development. *Cell*, **44**(2), 365–72.

Elder K, Johnson MH (2015c). The Oldham Notebooks: an analysis of the development of IVF 1969–1978. iii. Variations in procedures (c). *Reprod Biomed and Society, in press.*

Guérin P, Ménézo Y (1995). Hypotaurine and taurine in gamete and embryo environments: de novo synthesis via the cysteine sulfinic acid pathway in oviduct cells. *Zygote*, **3**, 333–43.

Guérin P, Rosset E, Rey M, *et al.* (2012). Amino acids in cat fallopian tube and follicular fluids. *Theriogenology*, **77**, 558–62.

Hammamah S, Ménézo Y (1999). *Ovocyte et embryon: De la physiologie à la pathologie.* Ed Elipses, Paris, p. 204.

Heo YS, Cabrera LM, Song JW, *et al.* (2007). Characterization and resolution of evaporation-mediated osmolality shifts that constrain microfluidic cell culture in poly (dimethylsiloxane) devices. *Anal Chem*, **79**, 1126–34.

Lawitts JA, Biggers JD (1991a). Optimization of mouse embryo culture media using simplex methods. *J Reprod Fertil*, **91**, 543–6.

Lawitts JA, Biggers JD (1991b). Overcoming the 2-cell block by modifying standard components in a mouse embryo culture medium. *Biol Reprod* **45**, 245–51.

Ménézo Y, Laviolette P (1972). Amino constituents of tubal secretions in the rabbit. *Ann Biol Anim Biochim Biophys*, **12**(3), 383–96.

Ménézo Y, Guerin P (1997). The mammalian oviduct: biochemistry and physiology. *Eur J Obstet Gynecol Reprod Biol*, **73**(1), 99–104. Review.

Quinn P, Kerin JF, Warnes GM (1985). Improved pregnancy rate in human in vitro fertilization with the use of a medium based on the composition of human tubal fluid. *Fertil Steril* **44**(4), 493–8.

Swain JE, Cabrera L, Smith GD (2012). Microdrop preparation factors influence culture media osmolality which can impair mouse preimplantation embryo development. *RBM Online*, **24**(2), 142–7.

Swain JE, Cabrera L, Xu X, Smith GD (2010). Environmental factors and manual manipulations during preparation influence embryo culture media osmolality. *Fertil Steril*, **94**(4), S32.

Wang F, Kooistra M, Lee M, Liu L, Baltz JM (2011). Mouse embryos stressed by physiological levels of osmolarity become arrested in the late 2-cell stage before entry into M phase. *Biol Reprod*, **85** (4), 702–13.

Whitten WK, Biggers JD (1968). Complete development in vitro of the pre-implantation stages of the mouse in a simple chemically defined medium. *J Reprod Fertil*, **17**(2), 399–401.

Whittingham DG (1971). Culture of mouse ova. *J Reprod Fertil (Suppl)* **14**, 7–21.

Chapter 5

pH regulation in gamete and embryo culture systems

Introduction: intracellular pH homeostasis

Correct development of the early preimplantation embryo requires a complex and carefully orchestrated series of essential metabolic reactions. These reactions are carried out by enzymes, whose correct function is highly pH dependent. Intracellular pH (pHi) regulates protein conformation, glycolysis, cytoskeletal dynamics, and a number of other critical metabolic, cell transport, and epigenetic functions (Swain & Pool, 2009). Minor environmental changes that affect the extracellular pH (pHe) can have a profound impact on pHi, and affect significant cell functions with inevitable effects on both fertilization and successful embryo development.

At early cleavage stages, the embryo lacks key regulatory functions to control pHi and osmotic pressure. Pyruvate is the primary energy source, metabolized via mitochondrial oxidative phosphorylation. In order to produce ATP through ATP synthetase, mitochondria need a proton gradient, with a high hydrogen ion (H^+) concentration at the space between inner and outer membranes, and a low H^+ concentration in the matrix. H^+ can pass through the outer mitochondrial membrane, but the inner membrane is impermeable to H^+. When mitochondria are exposed to a low pH environment, the resulting proton gradient allows generation of ATP through ATP synthetase (Cooper & Hausman, 2013).

After compaction, following activation of the embryonic genome and its transcription to provide more complex functional cellular mechanisms, embryos require rapid ATP production. In most mammalian species, metabolism shifts to glycolytic energy production via the Emden–Meyerhof–Parnas (EMP) pathway, with glucose as the primary substrate. pHi is an important regulator for this pathway, through key enzymes such as phosphofructokinase (PFK), which is active only over a narrow pH range. Lowering of pHi will decrease glycolytic activity and affect the supply of ATP needed to fuel numerous metabolic functions at this critical stage.

Since pHi is influenced by pHe, maintaining stable and appropriate pH of the culture medium is critical in order to prevent perturbations in crucial metabolic pathways. Table 5.1 outlines some of the effects that may result from inadequate pH control; these effects are due to disruption of basic biochemical metabolic pathways that depend on ATP supplied by correctly functioning mitochondria.

Substrate and oxygen availability also affect pHi, shifting metabolic pathways toward routes that influence ATP production. The biochemical pathways that make up the tricarboxylic acid (TCA) cycle (respiration) utilize substrates at a much slower rate compared with those involved in glycolysis (fermentation). These effects were originally described in

Table 5.1 Cytoskeletal and metabolic effects of pH variation

Effects of decreased pH	Effects of elevated pH
Disrupted localization of mitochondrial filaments	Increased glycolysis through the phosphofructokinase pathway
Disrupted localization of actin filaments	Lowered oxidative metabolism

the yeast *Saccharomyces cerevisiae*, a facultative anaerobe that can produce energy using different metabolic pathways according to the availability of oxygen (O_2):

1. *Crabtree effect* (Lehninger, 2008): inhibition of respiration (O_2 consumption) by glycolysis.
   In the presence of high external glucose concentrations, metabolism is shunted to produce ATP via increased glycolysis and substrate-level phosphorylation, instead of oxidative phosphorylation through the mitochondrial electron transport chain and TCA cycle. This shift results in decreased O_2 consumption, and in yeast, ethanol and CO_2 are produced. Increased glycolysis requires inorganic phosphate and the coenzyme nicotinamide adenine dinucleotide (NAD), and this limits their availability for phosphorylative oxidation in the TCA cycle.
2. *Pasteur effect* (Krebs, 1972): inhibition of fermentation by O_2, so that there is a decrease in carbohydrate breakdown when conditions switch from anaerobic to aerobic conditions.
   In low O_2 concentrations, the product of glucose metabolism (pyruvate) is rapidly metabolized to ethanol and CO_2 with low energy production (2 moles of ATP per mole of glucose). With increased availability of O_2, pyruvate is converted to acetyl coenzyme A (acetyl CoA) that can be used in the TCA cycle, producing 34 moles of ATP per mole of glucose (slow aerobic metabolism).Therefore, the same amount of ATP requires about 15 times more glucose under low O_2 concentrations. With increased availability of O_2, ATP production increases; this inhibits glycolysis, because ATP acts as an allosteric inhibitor for phosphofructokinase, an enzyme in the glycolytic pathway.

Physiologically, this shift from slow aerobic to rapid anaerobic metabolism occurs whenever mitochondria are deprived of O_2, as happens with strenuous exercise: mitochondria produce ATP by consuming glucose faster via the less efficient fermentation to lactate, which accumulates in muscle tissue.

Although these elegant biochemical studies in yeast demonstrated the complexity and potential synergy/feedback between different metabolic pathways, the situation in human gametes and embryos is far more complex (Quinn & Cooke, 2014). Interactions between different classes of metabolites (sugars, lipids, amino acids) are not necessarily the same in differing systems. *In vitro*, the "free floating" human embryo must achieve the correct balance between active uptake of sugars, amino acids, and nucleic acid precursors, as well as passive transport of lipids, mainly associated with albumin. The culture medium must be equilibrated and make specific metabolites available in the appropriate ratios (Ménézo *et al.*, 2013).

Intracellular pH (pHi) in human gametes and embryos has been examined at different stages of development. Dale *et al.* (1998) reported a range from pH 7.3 ± 0.3 in immature germinal vesicle (GV) stage oocytes, to pH 7.5 ± 0.2 in aged Metaphase II oocytes. Phillips *et al.* (2000) reported a slightly lower pH range, from pH 7.04 ± 0.07 in immature GV stage oocytes to pH 6.98 ± 0.2 in fresh Metaphase II oocytes, as illustrated in Table 5.2.

Phillips *et al.* (2000) also demonstrated that the mean pHi in early human cleavage stage (2–8 cell) embryos remains relatively constant at pH 7.12 ± 0.01, regulated by three

Table 5.2 Intracellular pH (pHi) measured in human oocytes and embryos

Cell stage	pHi (Philips *et al.* 2000)	pHi (Dale *et al.* 1998)
Prophase I (Germinal vesicle)	7.04 ± 0.07 (*n*=4)	7.3 ± 0.3 (*n*=13)
Metaphase I oocyte	7.03 ± 0.04 (*n*=9)	N.A.
Metaphase II oocyte (fresh)	6.98 ± 0.02 (*n*=9)	7.4 ± 0.1 (*n*=22)
Metaphase II oocyte (aged)	N.A.	7.5 ± 0.2 (*n*=26)
Fertilized oocyte (2 PN)	N.A.	7.4 ± 0.1 (*n*=18)
Two–eight-cell embryo	7.12 ± 0.01 (*n*=199)	N.A.

Table 5.3 Interspecies variation in pHi of oocytes and embryos at different stages of development (adapted from Swain, 2012)

Species	Cell type	pH	Reference
Mouse	1-cell	≈7.1	Zhao *et al.* (1995,1996)
	2-cell	≈7.1	
	Morula	≈7.1	
	Blastocyst	≈7.1	
	Zygote	7.17	Edwards *et al.* (1998)
	2-cell	7.19	
	8–16 cell	7.21	
	Morula	7.22	
	GV-intact	≈7.11	Fitzharris & Baltz (2006)
	Follicle enclosed	≈7.28	
	M II	7.13	House (1994)
	GV	6.96	Phillips & Baltz (1996)
	M II	7.00	
	Fertilized	7.02	
Hamster	1-cell	7.19	Lane *et al.* (1998)
	2-cell	7.21	
	3-cell	7.22	
Human	GV-intact M I	7.04	Phillips *et al.* (2000)
	M II	7.03	
	Cleavage	6.98-7.12	
Cow	M II	7.2	Lane & Bavister (1999)
	2-cell	7.24	
	4-cell	7.24	
	8-cell	7.20	

ion-exchange/antiporter mechanisms that compensate for alkalosis/acidosis in order to maintain pHi within a range of pH 7.0 – 7.3 (see below). Edwards *et al.* (1998) suggested that the ability to regulate pHi differs at different embryonic stages: morula and blastocyst stage mouse embryos can regulate their pHi better than embryos at earlier cleavage stages. Table 5.3 demonstrates interspecies variation in the pHi of oocytes and embryos at different stages of development (adapted from Swain, 2012).

The cell cytoplasm and its associated proteins provide a limited and short-term regulation of pHi. Long-term regulation depends upon export of H^+ and import of HCO_3^- across the cell membrane, and this is carried out chiefly by three membrane-bound molecules: Na^+/H^+ antiporter, H^+ monocarboxylate co-transporter, and HCO_3^-/Cl^- transporter. The Na^+/H^+ antiporter exists in five different isoforms, and each isoform differs in kinetic and tissue specificity. The NHE1 (Na^+/H^+ exchanger) isoform can be activated by growth factors via phosphorylation; mRNA for this isoform has been detected in the mouse oocyte and at all stages up to the blastocyst (Barr *et al.*, 1998; Lane *et al.*, 1998, 1999a, 1999b).

In vivo, endogenous pHe is regulated by a bicarbonate buffered system via carbonic anhydrase and circulating CO_2; *in vitro*, the pH of culture medium (pHe) is determined by the level of CO_2 in the gassed atmosphere and the amount of CO_2 associated with bicarbonate in the culture medium. However, culture media formulations cannot take into account the efficacy of transport systems present in the oocyte and early embryo for handling other substrates that influence pHi. Studies carried out in hamster embryos showed that the early embryo has very little capacity to regulate pHi, and deviations of only 0.2 pH units can have a detrimental effect on subsequent development (Zander-Fox *et al.*, 2010).

Amino acids in culture media act as organic osmolytes (Dawson & Baltz, 1997; Baltz, 2012), and also play a role in pHi regulation. This was proposed by Bavister's hypothetical model, whereby glycine and/or taurine shuttle across the cell membrane with Na^+ salt (influx) and H^+ as free acid (efflux), thereby exporting protons and buffering the pHi to prevent intracellular acidosis (Bavister, 1995). Clearly, careful regulation of ionic homeostasis is important to avoid external pH fluctuations (Fig. 5.1).

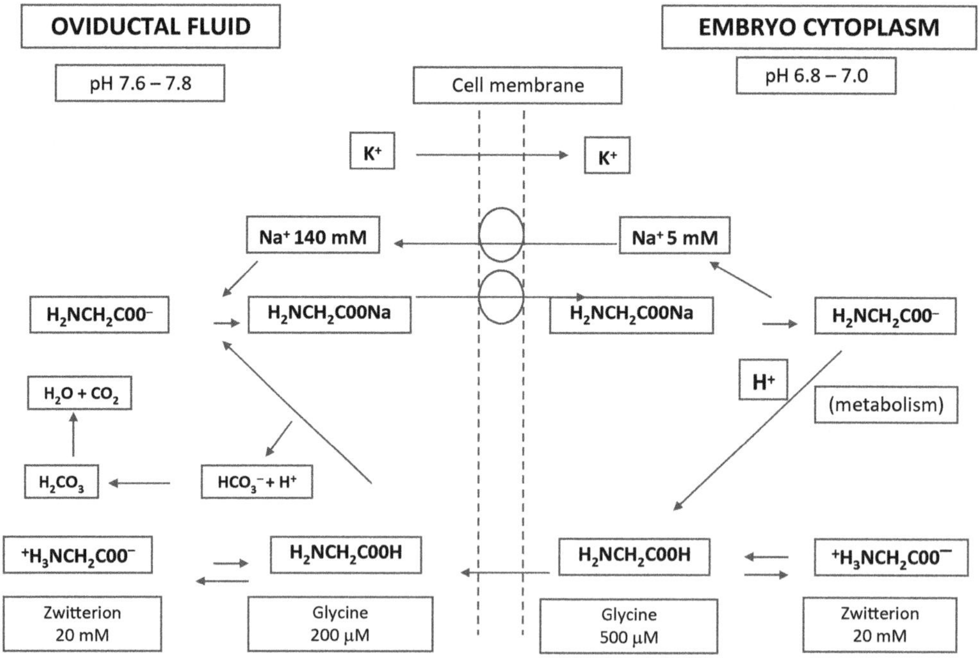

Fig. 5.1 Bavister's model for regulation of intracellular pH via amino acid shuttling across the cell membrane. *(Adapted with permission from: Bavister BD (1995). Hum Reprod Update* ***1(2)****: 91–148.)*

Table 5.4 Inter-species variation in external pH values (pHe) of follicular fluid, oviductal fluid, and uterine secretions *(adapted from* Swain, *2012, with permission)*

Species	Site	pHe	Reference
Cow	Oviduct	7.60	Hugentobler *et al.* (2004)
		6.4–6.7	Olds & VanDemark (1957)
	Uterus	6.96	Hugentobler *et al.* (2004)
Monkey	Oviduct, follicular phase	7.2	Maas *et al.* (1977)
	Oviduct, ovulation	7.6	
Rabbit	Oviduct	7.8–8.2	Hamner & Williams (1965)
		7.9	Vishwakarma (1962)
	Estrous	7.8	Iritani *et al.* (1971)
	Pseudo-pregnant	7.4	
	Isthmus	~7.24	Beier (1974)
	Infundibulum	~7.67	
	Uterus	7.9	Iritani *et al.* (1969)
		~7.4–7.8	Beier (1974)
		7.8	Vishwakarma (1962)
Sheep	Oviduct	7.4	Iritani *et al.* (1969)
		6.8–7.0	Hadek (1953)
	Uterus	7.0	Iritani *et al.* (1969)
Human	Follicle	7.27	Shalgi *et al.* (1972)
		7.34	Fraser *et al.* (1973)
		7.26	Imoedemhe *et al.* (1993)
	Uterus	6.5–6.7	Sedlis *et al.* (1967)
		7.12	Yedwab *et al.* (1976)

The pHe that is considered to be "optimal" for *in vitro* culture of gametes and embryos is based upon numerous variables and factors: not only comparison of *in vivo* vs. *in vitro* environment, but also pHi and its effect upon essential biochemical functions of the cells. The pH of human body fluids has an approximate range of pH 7.2 to 7.5, and the pH of oviductal and intrauterine fluids may vary at different stages of the menstrual cycle and at different locations from proximal to distal sections of the Fallopian tube (Fisch *et al.*, 1990). Some of these differences are species dependent, as shown in Table 5.4. This is the basis for the slight variation in pH of some commercially available stage-specific sequential culture media, with early stage media having a slightly higher pH. However, oocyte sibling studies comparing sequential vs. single-stage media between Day 1 and Day 5 have shown that the overall clinical and embryological parameters did not differ significantly with the use of either single-stage or sequential media (Çiray *et al.*, 2012).

Box 5.1 Origin of the term "pH"

The concept of pH scale was first described in 1909, at the Carlsberg Laboratory in Copenhagen. Whilst studying the effect of ion concentrations on proteins, the Danish biochemist Søren Peder Lauritz Sørensen realized that the concentration of hydrogen ions was of crucial importance, and introduced a scale as an indirect measure of acidity, represented by the negative logarithm of hydrogen ion concentration: pH = $-\log [H^+]$.

"The magnitude of the hydrogen ion concentration will accordingly be represented by means of the normality factor with regard to the hydrogen ion, and this factor will be written in the form of a negative potenz (power) of 10. . .I employ the name "hydrogen ion exponent" and the symbol pH for the numerical value of this potenz (power). . ." (S. Sørensen (1909). Enzymstudien II: Uber die Messung und die Bedeutung der Wasserstoffenenkonzentation bei enzymatischen Prozessen. *Biochem Zeit*, **21**, 131–200)

Suppliers of commercial media generally recommend a pHe range of 7.2 – 7.4 for embryo culture (Quinn & Cooke, 2004). However, bearing in mind that the pH scale is logarithmic, a 0.2 variation in pHe represents a 60% change in H^+ concentration! Thus, it is important to keep the pHe range as narrow as possible in order to reduce variability in the culture system. As described below, medium pH is affected by the overall stability of the culture system, and factors such as temperature, humidity, altitude, incubator gas concentration, and phase, as well as correct handling, will affect the pHe range; a working knowledge of how pH is determined, regulated, and measured is an essential foundation for troubleshooting an IVF culture system (Box 5.1).

Physico-chemical regulation of pH

pH measures the molar concentration of hydrogen ions $[H^+]$ in a solution, given by the formula pH = $-\log [H^+]$.

A solution with $[H^+]$ of 0.0000001M will have pH = 7

Adding hydrogen ions to a solution causes it to become more acidic (lower pH), and their removal makes it more alkaline or basic (higher pH). The pH in bicarbonate buffered systems is determined by the Henderson–Hasselbalch (H–H) equation. pH is related to an equilibrium between gas phase CO_2 and CO_2/HCO_3 dissolved in the medium, with carbonic acid as an intermediate:

$$pH = pK_a + \log_{10} [H_+]$$

where,

$\log_{10}[H^+]$ = log of the reciprocal of the molar concentration of H^+ ions
K_a = the acid dissociation constant of a solution, i.e. the point where equal portions of acid and base exist in solutions and equilibrium is reached
$pK_a = -\log_{10} K_a$, the ionisation constant for the acid

(lower pK_a values = stronger acids)

For a bicarbonate buffered system

pK_a = 6.1
molar concentration of H^+ ions = 52 mg/ml

Table 5.5 Effect of temperature at different ranges of pH

Temperature °C	pH 4.00 ± 0.01*	pH 7.00 ± 0.01*	pH 10.00 ± 0.01*
10	4.00	7.07	10.18
15	4.00	7.04	10.14
20	4.00	7.02	10.06
25	4.00	7.00	10.00
30	4.01	6.99	9.95
35	4.02	6.98	9.91
40	4.03	6.97	9.85
50	4.05	6.96	9.78
60	4.08	6.96	9.75

(*) = values at 25 °C.

Table 5.6 pH variation according to $NaHCO_3$ and CO_2 concentrations

g/L $NaHCO_3$	5% CO_2	5.5% CO_2	6% CO_2	6.5% CO_2
1	7.07	7.03	6.98	6.95
1.2	7.16	7.11	7.07	7.03
1.4	7.23	7.19	7.15	7.11
1.6	7.29	7.25	7.21	7.17
1.8	7.35	7.30	7.26	7.23
2	7.4	7.35	7.31	7.28
2.2	7.44	7.40	7.36	7.32
2.4	7.48	7.44	7.40	7.36

In practical terms, the H–H equation for a bicarbonate-buffered medium is

$$pH = 6.1 + \log (52\ [mg/ml\ NaHCO_3\ /\%CO_2]-1$$

The resulting actual pH is therefore affected by:

- The presence of other solutes such as amino acids and complex salt mixtures, since pK_a is a function of salt concentration.
- Temperature: a good pH meter should have a probe that compensates for temperature. Increasing temperature enhances solubility and ion dissociation, liberating protons (H^+) and thus decreasing pH. Temperature increase also has a de-gassing effect, so that CO_2 in bicarbonate-buffered culture media is released, increasing pH. The effect of temperature differs at different ranges of pH, as illustrated in Table 5.5.

Table 5.6 illustrates the range of variation in pH according to $NaHCO_3$ and CO_2 concentrations; exchange between gaseous and dissolved CO_2 means that the buffer

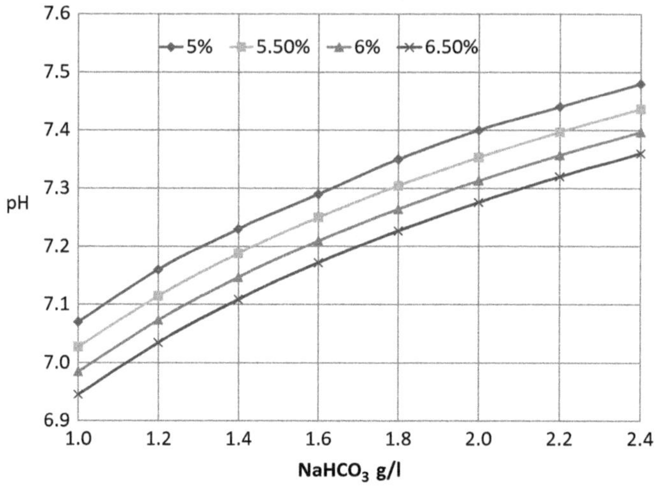

Fig. 5.2 Nomogram for estimating a pH value at different CO_2 and $NaHCO_3$ concentrations.

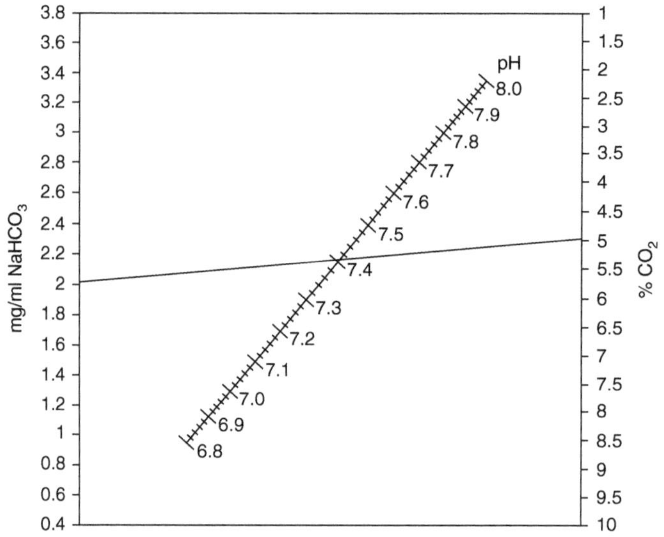

Fig. 5.3 Nomogram for estimating % CO_2 at different pH values, according to bicarbonate concentration.

equilibrium in the aqueous phase is dependent upon CO_2 concentration in the gas phase, according to the equation:

$$CO_2 \text{ (gas)} \leftrightarrow CO_2 \text{ (dissolved)} + H_2O \leftrightarrow H_2CO_3 \leftrightarrow H^+ + HCO_3$$

Figures 5.2 and 5.3 illustrate examples of nomograms that can be used to estimate corresponding $NaHCO_3$, CO_2, and pH values; Figure 5.4 demonstrates the influence of CO_2 concentration on pH in media that contain different concentrations of $NaHCO_3$.

Theoretically, the pH in bicarbonate-buffered systems is also dependent upon the partial CO_2 pressure, which is influenced by altitude. Media pH is determined by the amount of CO_2 present in the incubator atmosphere, and increasing the % CO_2 may be considered if the laboratory is in a mountain region (Table 5.7). However, the practical calculations below demonstrate that the real effect is minimal: an elevation of 2000 m will result in an increase of 0.13 pH units.

Table 5.7 Variation in pH with altitude/ partial CO_2 pressure. An increase of 2400 meters in altitude results in an increase of 0.13 pH units at a constant 5% CO_2 concentration *(Quartec GmBH)*.

Altitude	Atm. press.	pCO 5%	(0.03 × pCO_2)	$\log_{10}[(HCO_3^-)/(0.03 \times pCO_2)]$
0 m	760 mm Hg	0.05 × 760 = 38	= 1.14	$Log_{10}(25/1.14 = 21.9) = 1.34$
400 m	720 mm Hg	0.05 × 720 = 36	= 1.08	$Log_{10}(25/1.08 = 23.1) = 1.36$
2438 m	568 mm Hg	0.05 × 568 = 28	= 0.85	$Log_{10}(25/0.85 = 29.4) = 1.47$

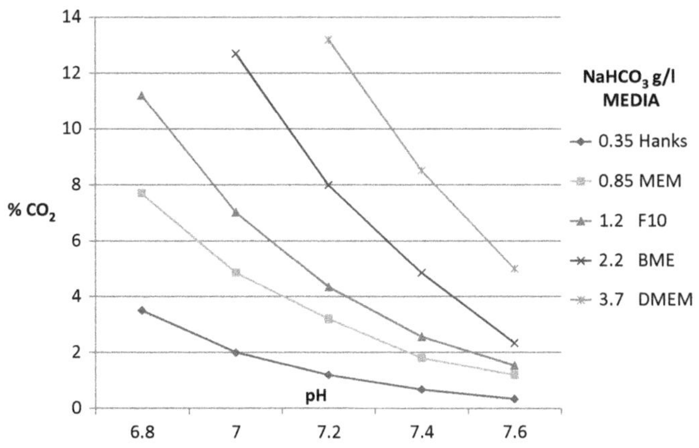

Fig. 5.4 pH variation in different culture media according to CO_2 concentration.

$$pH = pK_s + \log_{10} \frac{c(A^-)}{c(HA)}$$

$$pH = pK_a + \log_{10}[HCO_3^-/(H^+)]$$

$$pH = pK_a + \log_{10}[HCO_3^-/(0.03 \times pCO_2)]$$

(0.03 = solubility of CO_2 mm/l/mmHg)

An IVF laboratory at 3000 feet above sea level can maintain pH 7.2 using 6% CO_2, whilst a laboratory at around 10000 feet above sea level may require a higher level for similar pH; the incubator display shows the correct % CO_2, but lower partial pressure means that this percentage represents fewer CO_2 molecules in the atmosphere (Conaghan, 2014). However, other environmental factors such as humidity also need to be considered. If a thermal conductivity (TC) sensor is used to monitor CO_2 levels (see Chapter 8), incubator humidity may need to be increased if the external environment is very dry (extra pans of sterile water) to ensure that the sensors are functioning correctly. By contrast, humidity of the air within the laboratory may need to be reduced if the external environment is very humid; this is also important in reducing the risk of fungal infection (see Chapter 9).

Culture media pH equilibration

As with osmolality (Chapter 4) and temperature (Chapter 6), pH equilibration in media is directly related to the culture system that is used, most importantly to the surface area: volume ratio of solutions in the dishes, as well as use of lids and oil overlays. Dishes that are used without an oil overlay are extremely sensitive to pH changes when placed in, or removed from an incubator: 1 ml of medium in a four-well dish will reach a stable pH after around 2 hrs in a 5% CO_2 in air atmosphere, whereas 0.5 ml will equilibrate more quickly under the same conditions. Conversely, when a dish with 1 ml of medium is removed from the incubator and kept at 37 °C, the pH will begin to rise to non-physiological levels immediately: 1 ml of medium will reach pH of 7.52 after 5 minutes in room air, and 0.5 ml can reach pH 7.77 under the same conditions (Conaghan, 2014). Using an oil overlay can dampen dramatic pH shifts, but this also dramatically increases the time required for initial equilibration in a CO_2 atmosphere. CO_2 diffuses very slowly through oil, and 0.5 ml of medium with 0.5 ml oil overlay takes almost 12 hrs to reach stable pH in the incubator. A 50 μl droplet under 0.5 ml of oil takes around 8 hrs to equilibrate, and will maintain a stable pH for at least 5–10 minutes in room air at 37 °C. Longer time periods will cause greater pH excursions, which will not be immediately apparent due to the slow diffusion of gas through the oil; re-equilibration of pH in the incubator after 30 minutes' exposure to room air takes over 2 hrs.

Buffers and zwitterions

Familiarity with the properties of different buffers is essential in the IVF laboratory, particularly if a culture system is changed to one that involves the use of a different buffer (Palasz *et al.*, 2008; Swain, 2010; Will *et al.*, 2011). Zwitterions are large organic amino acids that can behave either as acids or bases, with a wide range of buffering capacity. Zwitterion pK_a is close to the working pHe of culture media, and they maintain a stable pH in air. During the 1960s, Norman Good and colleagues at Michigan State University prepared and described a number of buffering agents that fulfilled specific criteria relevant to biological systems: pK_a between 6.0 and 8.0, non-toxic, high solubility, low permeability to biological membranes and limited effect on biochemical reactions, minimal changes with temperature/concentration (Good *et al.*, 1966). These buffers are known as 'Good's Buffers'. Three examples of Good's Buffers that are used in IVF culture systems are:

- HEPES: 4-(2-hydroxyethyl)-1piperazineethanesulfonic acid
- MOPS: 3-(N-morpholino-)-propanesulfonic acid
- DIPSO: 3-(N,N-bis[2-hydroxyethyl]amino)-2-hydroxypropanesulfonic acid

The pK_a of these buffers is highly temperature-dependent, as shown in Table 5.8.

Note that MOPS buffer has the lowest pK_a, and is therefore a more effective buffer for use in a lower (acidic) pH range (Table 5.8).

Maintaining stable pH during manipulations outside the incubator is crucial to embryo development, and therefore a zwitterion-containing buffer is usually used for gamete and embryo handling in room atmosphere: follicle flushing, oocyte washing, cryopreservation protocols, sperm preparation, micromanipulation (ICSI, assisted hatching, biopsy, etc). HEPES-buffered medium is equilibrated to a pH of 7.4 in the presence of bicarbonate, and maintains a relatively stable pH even in ambient air.

Table 5.8 pKa and effective pH range of zwitterion buffers at 24 °C and 37 °C.

Name	pKa 24 °C	pKa 37 °C	Buffer range
HEPES	7.48	7.31	6.8 – 8.2
*MOPS	7.2	7.02	6.5 – 7.9
DIPSO	7.60	7.35	7.0 – 8.2

* MOPS has a lower initial pKa, and therefore is more suitable than the other two buffers for use in acidic pH range.

Important notes: Exposure to a higher than normal CO_2 atmosphere (such as 5% CO_2) will lower the pH of HEPES buffered medium, and it will become acidic. Therefore, where HEPES is used for oocyte collections, ICSI, or any other "out of the incubator" procedures:

1. Do not expose the HEPES-buffered medium to 5% CO_2 conditions, i.e. don't prepare dishes or tubes with loose lids which contain HEPES-buffered medium and place them into 5% CO_2 incubators at 37 °C to warm them.
2. Pre-equilibrate HEPES-buffered medium to the correct temperature in tight-lidded dishes or tubes only.
3. Prepare the loose-lidded dishes just prior to the procedure:

   If loose-lid dishes are prepared with HEPES-buffered medium several hours before a micromanipulation procedure is performed, the oocytes or embryos will be transferred into media that has become acidic. Do not do this! The problem can be overcome by replenishing droplets from a stock of warmed HEPES-buffered medium held in a sealed tube in the same incubator.
4. Do not use HEPES-buffered media for micromanipulation in a sealed-glove access system with an atmosphere of 5%–6% CO_2.
5. Wash gametes and embryos thoroughly in HEPES-free culture medium before transferring into the same HEPES-free culture medium in the CO_2 culture incubator. This includes after oocyte collection, after ICSI, and after embryo biopsy.
6. A shift between bicarbonate and HEPES buffers may cause some metabolic stress, with the result that metabolism is diverted to different metabolic pathways. In mouse embryos, Lane & Gardner (1996) demonstrated that:

   "culture of *in vivo* developed mouse blastocysts in a simple culture medium based on a balanced salt solution supplemented with carbohydrates for 3h significantly perturbs the metabolism of both mouse and rat embryos."

   inducing blastocysts to increased glycolytic activity. The developmental competence of mouse blastocysts after transfer was significantly reduced after just 6h of culture in a simple medium. However, the presence of amino acids or vitamins in the culture medium decreased both the glycolytic activity and the oxidative capacity of cultured mouse blastocysts, and apparently compensated for the loss in viability.
7. If the HEPES-buffered medium or PBS contains no amino acids or energy substrates, then handling should be restricted to very short term only to avoid compromise to gamete or embryo health.

   Detrimental effects of prolonged exposure to HEPES have been reported (Morgia *et al.*, 2006), but the real effects have been questioned, with the suggestion that the results may be species dependent, as well as a reflection of HEPES interactions with other components in the culture system rather than direct toxicity of the buffer itself (Zigler *et al.*, 1985; Swain & Pool, 2009). Nonetheless, oocytes, and in particular denuded oocytes, are highly sensitive to changes in pH (Butler *et al.*, 1988; Downs & Mastropolo, 1997) and micromanipulations should be carried out as quickly as possible. For example, consider the number of oocytes that will be placed in the dish before preparing an ICSI dish with HEPES-buffered medium, so that the rest of the oocytes awaiting ICSI are kept in the buffer for a minimal amount of time. The number of oocytes per ICSI dish varies between clinics, according to the skill and speed of the ICSI practitioner (see Chapter 11). As an alternative option, ICSI can be performed in pre-equilibrated culture medium without HEPES, using droplets of the standard formulation used in the laboratory incubators. However, in this case micromanipulation should be limited to three oocytes only, or to keeping the dish out of the incubator for a maximum of 5 minutes to minimize the risk of a pH change. It is also advisable to have the CO_2 incubator located close to the micromanipulation station if using medium without HEPES buffer.

Buffers that use a combination of HEPES, MOPS, and DIPSO in a single medium have been suggested as a means of reducing the individual zwitterion concentration and providing optimal pH buffering capacity for IVF handling media. Combining buffers not only reduces the absolute concentration of individual components so that concerns about potential toxicity are reduced, but also allows the buffering capacity to be expanded over a range of temperatures (Swain & Pool, 2009; Will & Swain, 2012).

Measuring pH

Although pH is a critical parameter, in practice it is not easy to monitor effectively. A standard glass probe pH meter can be used for solutions, but it cannot measure pH under culture conditions such as microdroplets under oil. However, new batches of medium can be tested in a larger (e.g. 2 ml) volume after overnight equilibration in standard culture conditions; these readings can be used to keep a QC log of media pH by batch (Conaghan, 2014).

A number of alternative devices are available for use in IVF culture systems, but different devices can yield different results when used to test various media. Therefore, whatever device is used, careful validation and calibration is essential; it is unwise to rely on only one method. pH measurement is also highly temperature dependent: calibration and sample measurement should be recorded, and both performed at the same temperature (Barron *et al.*, 2006). ISO 15189 ("Medical laboratories – Requirements for quality and competence") includes a detailed section on "Equipment calibration and metrological traceability": any devices used to validate equipment must function optimally, including pH meters.

Sample pH is measured after two-point calibration, with one buffer in an acid range and the other in an alkaline range. The value of the pH to be measured should lie between those two reference values, i.e. standard buffers of 4, 7, and 10 for culture media with a working pH of 7.2 – 7.4. Modern pH meters operate with a calibration procedure that automatically recognizes the pH of the calibration buffer, and the value requires only slight adjustment. The two reference values are plotted on a graph, and the slope between the two values is used to obtain the target value by referring to the electrode manual. The electrode must be

Box 5.2 Case report

The majority of embryologists will admit that measuring medium pH (pHe) after appropriate overnight equilibration is a more direct, and preferable monitoring tool than is measuring correct incubator gas concentration. However, the following example illustrates that this is not necessarily the case.

An experienced embryologist preferred to measure and monitor medium pH, and adjusted the CO_2 concentration in order to obtain the desired medium pH of 7.35. Surprisingly, 8.5% CO_2 was required in order to reach this pH in a culture medium based on modified Earle's Balanced Salt Solution. The incubator maintenance sheet showed that it had been serviced 6 months earlier, and no problems were detected. Measuring the percentage CO_2 with a calibrated gas-analyzer confirmed that the set-point, the display, and the obtained value for CO_2 all corresponded, and confirmed that the lab was effectively working with an atmosphere of 8.5% CO_2 for the culture system. Records also showed that the pH meter was calibrated before each measurement. However, the records mentioned only the operator and the data, without recording the pH value of calibration buffers, and with no indication of whether temperature correction was used. There was no record of the date of first use of the electrode, and no record of electrode performance.

An attempt to calibrate the pH meter with two buffers failed, and the meter reported an electrode problem. The pH of equilibrated medium was then measured using a blood gas analyzer, which confirmed a pH value of 7.22 after several measurements. It was later revealed that the electrode was at least 2 years old, that only one buffer was used for calibration, and that the users were not able to interpret the alarm warnings indicating that the electrode was exhausted [*"SPAN Error: A perfect pH electrode, at 25 °C produces 59.12 mV per pH unit. The difference between this perfect reading and the electrode's actual reading is called the span error. These theoretical values are not always achieved, even with brand new electrodes. New pH electrode performance specifications should meet the following criteria: TYPICAL SPECIFICATION – Offset: 7.00 +/−0.2 pH (+/−12 mV) SPAN: Better than 95% of theory; i.e. between 56.2 and 59.2 mV"* (Ross, 2004).]

Lessons to be learned:

1. pH measurement is considered to be one of the basic manipulations that laboratory personnel should be able to perform correctly; this is not necessarily the case in reality (see Conaghan, 2014, pp.145–6).
2. Users were not familiar with the operating instructions and were unable to use the instrument correctly. The instrument was purchased and used without training.
3. Users were not aware that electrodes do not last longer than 1 year.
4. Records were not sufficiently detailed.
5. Two buffers must be used for calibration, one above and one below the expected value.
6. Users were not aware that 8.5% CO_2 does not "fit" theoretically with the composition of their medium. They did not ask the company about the real composition of the medium, or about the correct percentage CO_2 to be used. This shows that information leaflets provided by companies are not always checked. Correct QM specifies that leaflets for each consumable must be checked: companies do change the version and also the content of these leaflets; the latest versions must be stored, with the date of validity (see Chapter 2).

This report shows the importance of having essential steps in QM: the "Plan" was excellent, the "Do" was done. . .but monitoring/checking was poor, and did not lead to appropriate action in order to correct the situation. An additional SOP about pH measurement, calibration, and pH-meter maintenance had to be written.

Box 5.3 **Practical tips for measuring pH**

Always read the manual/handbook, and follow the manufacturer's instructions with respect to handling, cleaning, storage, and calibration of the electrode.

1. Soak new electrodes in a 1:1 solution of saturated KCl and buffer at pH = 4.
2. The porous glass membrane (frit) must be free-flowing – keep it clean.
3. Always use fresh buffers (remember the influence of atmospheric CO_2).
4. Use temperature compensation and keep the temperature constant.
5. Mix/stir the sample whilst taking the reading.
6. Keep the filling hole on the reference electrode open whilst taking measurements.
7. Re-calibrate after 2 hours of use, or after 20–30 samples.
8. Use appropriate/matching electrical connections (BNC-BNC).

Box 5.4 **Case report**

Routine checks 1 hour before scheduled oocyte collection revealed that all of the culture media in the dishes held in the incubator was pink (alkaline), which indicated that the CO_2 level was too low. On checking the gas supply, the cylinder was empty (there was no alarm sensor on this incubator). The cylinder was replaced immediately to replenish the incubator CO_2 gas, and the team decided it was better to delay the oocyte collection procedure by 2 hours in order to re-equilibrate the media to the correct CO_2 level. Fortunately all of the follicles were still present, and the oocytes were then cultured in media at the correct pH.

Box 5.5 **Case report**

Routine checks prior to oocyte collection showed that the medium in culture dishes placed inside a benchtop incubator was yellow (acid). Further checks revealed that the benchtop incubator had been connected to a 100% CO_2 cylinder, instead of a 6% CO_2 cylinder. The 100% CO_2 cylinder was disconnected and replaced with a 6% CO_2 cylinder. Fortunately in this case, the large box incubator was correctly set up, with correctly equilibrated media, and it was not necessary to delay the oocyte collections. This incidence highlights the importance of knowing the difference between gas supplies required for box and benchtop incubators.

replaced if the measured and target values differ widely. Single-point calibration can be used as a rapid calibration method in between regular two-point calibrations, using a calibration buffer whose pH is close to the expected value of the solution being measured, usually pH 7.0 for culture media.

Albumin/proteins in culture media can cause a problem with pH measurements, because proteins tend to stick to the glass membranes of the electrode and will eventually block them. Electrodes can be cleaned with detergent and regenerated with the use of fluoric or hydrochloric acid, but this is a fastidious procedure that can potentially damage the electrode. As a first approach, the electrode can be heated in a 60 °C water bath for 30 minutes. However, regenerating and cleaning procedures are not always effective, and it is better to replace the electrode.

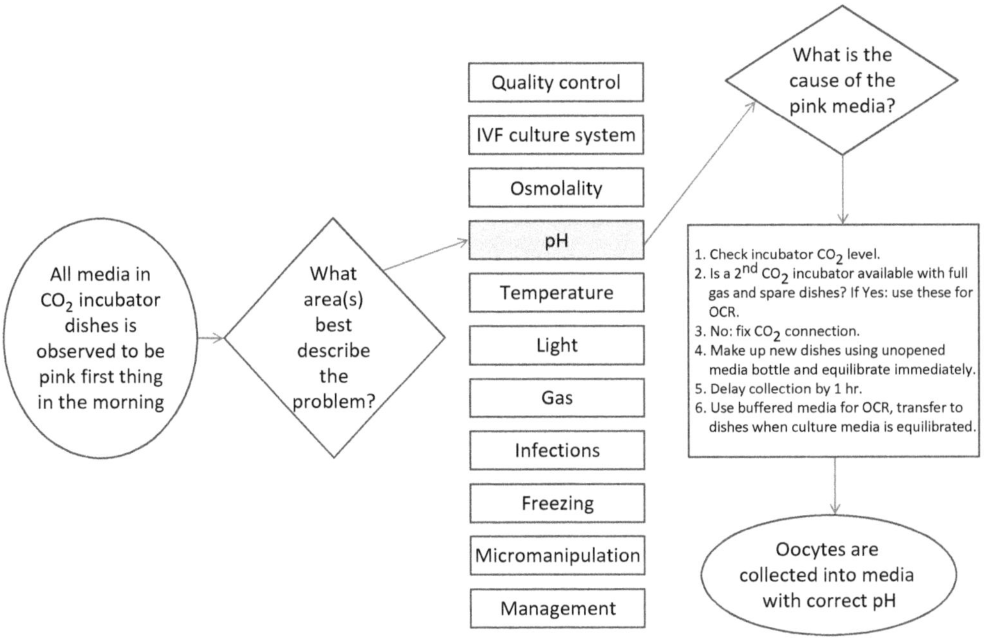

Fig. 5.5 Process map for troubleshooting pH problems.

Devices used to monitor pH in culture media

Devices available to monitor pH include "fluorescent decay time" pH meters and ion sensitive field effect transistor probes. The calibration method of each device should be evaluated (especially with regard to conformity to ISO15189) together with how easy the probe is to clean. Consideration should also be given to whether the device can measure pH in different culture systems – some probes are not suitable for measuring pH in microdroplets. pH meters can be used to provide single measures over a prolonged period (e.g. overnight incubation), or multiple probes can be used to provide continuous measures of media in multiple incubators simultaneously. The cost of different pH meters can vary considerably according to the device, and expense may also be a factor that affects the choice of pH meter.

Small bench-top incubators that use pre-mixed gas have been available since the mid-1990s; these smaller incubators have the advantage of rapid temperature and gas recovery after opening, but they use smaller gas tubing that may sometimes become blocked with condensed water. When this happens, accurate control of the gas mixture is impossible, due to lowered pressure and small reserve volume. pH can be tested in dishes in benchtop incubators, but as an alternative to measuring actual CO_2 level, gas levels are often validated by testing the purge rate of the incoming gas, which is usually set to around 15 ml/min.

A control "indicator dish" can be used to give an approximate ongoing assessment of microdroplet pH in a culture system, using culture medium containing the colored indicator phenol red (this is included as a constituent of some commercially available media or can be added as a supplement to other media). If the dish is placed on a sheet of white paper

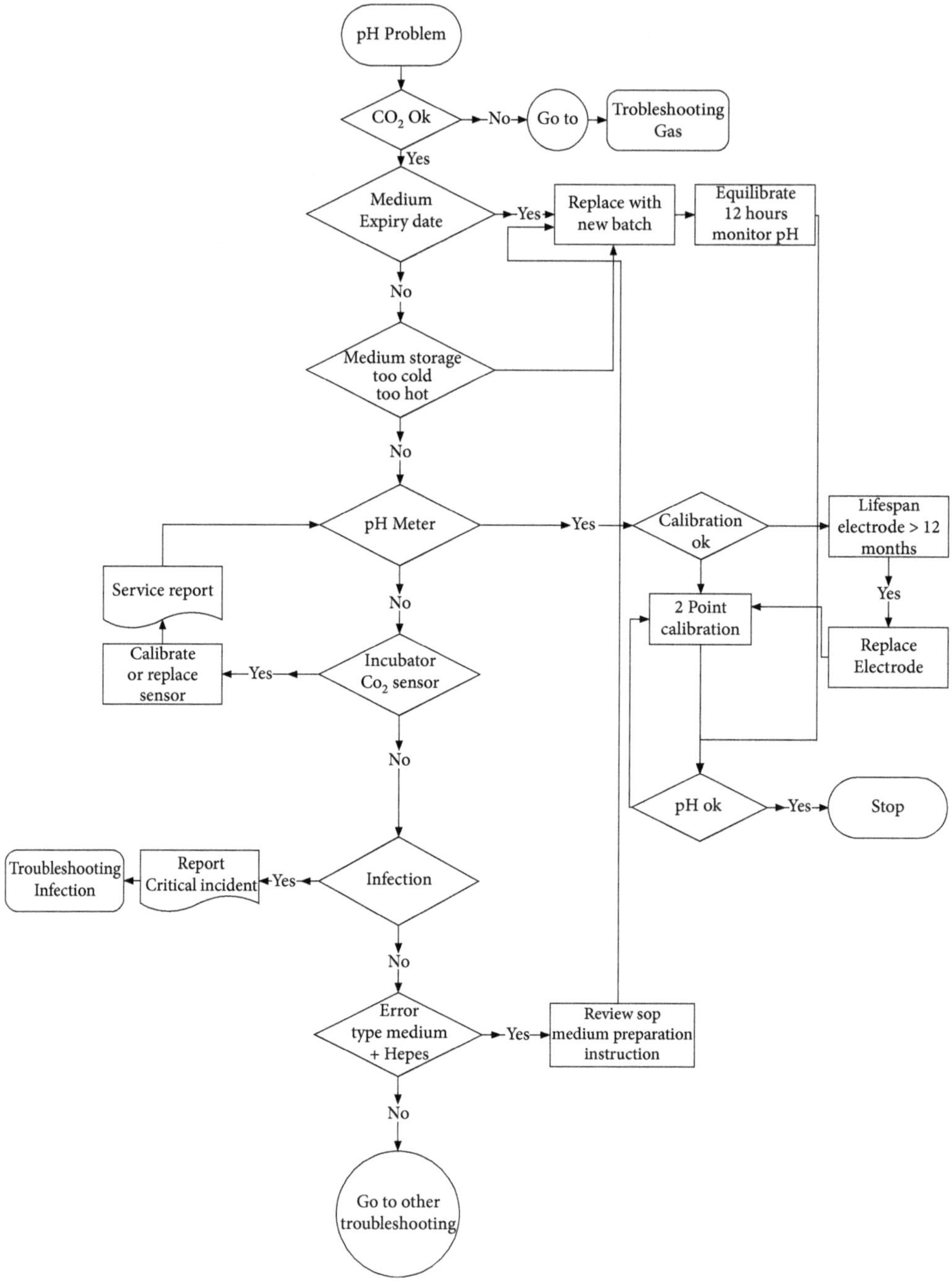

Fig. 5.6 Flowchart for troubleshooting pH problems.

in the culture incubator, the color of the media can be easily seen. The color can then be visually compared with the color of the pH reference media containing phenol red across the pH range: media turns yellow in acidic conditions pH<6.8 and pink in alkaline conditions pH >8.2.

A new system of measuring pH in IVF culture is now available from MTG (Bruckberg, Germany), which consists of a pH-sensitive proprietary disk inserted in one well of a four-well dish. pH readings are relayed continuously to a computer via optic fiber, recording real-time changes in medium pH.

Conclusion

A culture system must be carefully validated in order to confirm that the system functions at the optimal level: pH fluctuations are a sensitive indicator. Fluctuations within the incubator, gas phase, or culture system will be reflected in changes in pH, and measurement of pH over time and across all incubators is an important part of quality control monitoring in IVF systems. Although buffers have a role to play, they should be used only with active consideration of their differences and effects. Always bear in mind that a number of different factors can affect the measured pH in media, apart from measurement errors themselves: gas quality, pressure control/leaks in the gas delivery system, incubator temperature control, and quality/components of the media.

Further reading

Conaghan J (2014). pH control in the embryo culture environment. In *Culture Media, Solutions, and Systems in Human ART*, P. Quinn (ed). Cambridge University Press.

Lizotte M, Moeggenberg M (2013). Making good pH measurements http://www.ysi.com/media/support/YSI-Making-Good-pH-Measurements-web.pdf.

Ross M (2004). pH electrode performance. Sensorex Technical Bulletin. http://www.sensorex.com/support/more/ph_electrode_performance.

Swain JE (2010). Optimizing the culture environment in the IVF laboratory: impact of pH and buffer capacity on gamete and embryo quality. *RBM Online*, **21**(1), 6–16.

Swain JE (2012). Is there an optimum pH for culture media used in clinical IVF? *Hum Reprod Update*, **18**(3), 333–9.

Swain JE (2013). Comparison of three pH measuring devices within the IVF laboratory. *Fertil Steril*, **100**(3) S251.

References

Baltz JM (2012). Media composition: salts and osmolality. *Methods Mol Biol*, **912**, 61–80.

Barr KJ, Garrill A, Jones HD, Orlowski J, Kidder G (1998). Contributions of Na1/H1 exchanger isoforms to preimplantation development of the mouse. *Mol Reprod Dev*, **50**, 146–53.

Barron JJ, Ashton C, Geary L (2006). *The Effects of Temperature on pH Measurement.* Presented at the meeting of International Society of Electrochemistry, Edinburgh: www.reagecon.com, Technical papers.

Bavister BD, McKiernan SH (1993). Regulation of hamster embryo development *in vitro* by amino acids. In *Preimplantation Embryo Development*, BD Bavister (ed.) Springer-Verlag, New York pp. 57–72

Bavister BD (1995). Culture of preimplantation embryos: facts and artifacts. *Hum Reprod Update*, **1**(2), 91–148.

Beier HM (1974). Oviductal and uterine fluids. *J Reprod Fertil*, **37**, 221–37.

Butler JE, Lechene C, Biggers JD (1988). Noninvasive measurement of glucose uptake by two populations of murine embryos. *Biol Reprod*, **39**, 779–86.

Ciray HN, Aksoy T, Goktas C, Ozturk B, Bahceci M (2012). Time-lapse evaluation of human embryo development in single versus sequential culture media – a sibling oocyte study. *J Assist Reprod Genet*, **29**(9), 891–900.

Conaghan J (2014). pH control in the embryo culture environment. In *Culture Media, Solutions, and Systems in Human ART*, P. Quinn (ed). Cambridge University Press.

Cooper GM, Hausman RE (2013). *The Cell: A Molecular Approach* (Sixth Edition). ASM Press, Washington DC http://sites.sinauer.com/cooper6e/.

Dale B, Menezo Y, Cohen J *et al.* (1998). Intracellular pH regulation in the human oocyte. *Hum Reprod*, **13**, 964–70.

Dawson KM, Baltz JM (1997). Organic osmolytes and embryos: substrates of the Gly and beta transport systems protect mouse zygotes against the effects of raised osmolarity. *Biol Reprod*, **56**, 1550–8.

Downs SM, Mastropolo AM (1997). Culture conditions affect meiotic regulation in cumulus cell-enclosed mouse oocytes. *Mol Reprod Dev*, **46**(4), 551–66.

Edwards LJ, Williams DA, Gardner DK (1998). Intracellular pH of the mouse preimplantation embryo: amino acids act as buffers of intracellular pH. *Hum Reprod*, **13**, 3441–8.

Fisch B, Goldberg I, Ovadia J, Tadir Y (1990). Physicochemical properties of follicular fluid and their relation to *in vitro* fertilization (IVF) outcome. *JIVF-ET*, **7**(2), 67–73.

Fitzharris G, Baltz JM (2006). Granulosa cells regulate intracellular pH of the murine growing oocyte via gap junctions: development of independent homeostasis during oocyte growth. *Development*, **133**, 591–9.

FitzHarris G, Baltz JM (2009). Regulation of intracellular pH during oocyte growth and maturation in mammals. *Reproduction*, **138**(4), 619–27.

Fraser IS, Baird DT, Cockburn F (1973). Ovarian venous blood PO2, PCO2 and pH in women. *J Reprod Fertil*, **33**, 11–17.

Good NE, Winget GD, Winter W, *et al.* (1966). Hydrogen ion buffers for biological research. *Biochemistry*, **5**(2), 467–77.

Hadek R (1953). Alteration of pH in the sheep's oviduct. *Nature*, **171**(4361), 976.

Hamner CE, Williams WL (1965). Composition of rabbit oviduct secretions. *Fertil Steril*, **16**, 170–6.

House CR (1994). Confocal ratio-imaging of intracellular pH in unfertilised mouse oocytes. *Zygote*, **2**, 37–45.

Hugentobler S, Morris DG, Kane MT, Sreenan JM (2004). *In situ* oviduct and uterine pH in cattle. *Theriogenology*, **61**, 1419–27.

Imoedemhe DA, Chan RC, Ramadan IA, Sigue AB (1993). Changes in follicular fluid gas and pH during carbon dioxide pneumoperitoneum for laparoscopic aspiration and their effect on human oocyte fertilizability. *Fertil Steril*, **59**, 177–82.

Iritani A, Gomes WR, Vandemark NL (1969). Secretion rates and chemical composition of oviduct and uterine fluids in ewes. *Biol Reprod*, **1**, 72–6.

Iritani A, Nishikawa Y, Gomes WR, VanDemark NL (1971). Secretion rates and chemical composition of oviduct and uterine fluids in rabbits. *J Anim Sci*, **33**, 829–35.

ISO15189:2012 Medical Laboratories – Requirements for Quality and Competence.

Krebs H (1972). The Pasteur effect and the relations between respiration and fermentation. *Essays in Biochemistry* (8): 1–34.

Lane M, Bavister BD (1999). Regulation of intracellular pH in bovine oocytes and cleavage stage embryos. *Mol Reprod Dev*, **54**, 396–401.

Lane M, Baltz JM, Bavister BD (1998). Regulation of intracellular pH in hamster preimplantation embryos by the sodium hydrogen (Na^+/H^+) antiporter. *Biol Reprod*, **59**, 1483–90.

Lane M, Baltz JM, Bavister BD (1999a). Na^+/H^+ antiporter activity in hamster embryos is activated during fertilization. *Dev Biol*, **208**, 244–52.

Lane M, Baltz JM, Bavister BD (1999b). Bicarbonate/chloride exchange regulates intracellular pH of embryos but not oocytes of the hamster. *Biol Reprod*, **61**, 452–7.

Lane M, Bavister BD (1999). Regulation of intracellular pH in bovine oocytes and cleavage stage embryos. *Mol Reprod Dev*, **54**, 396–401.

Lane M, Gardner DK (1996). Selection of viable mouse blastocysts prior to transfer using a metabolic criterion. *Hum Reprod*, **11**, 1975–8.

Lane M, Gardner DK (2000). Regulation of ionic homeostasis by mammalian embryos. *Semin Reprod Med*, **18**, 195–204.

Lehninger A (2008). *Principles of Biochemistry.* W.H. Freeman & Co., New York, NY, p. 539.

Maas DH, Storey BT, Mastroianni L Jr. (1977). Hydrogen ion and carbon dioxide content of the oviductal fluid of the rhesus monkey (*Macaca mulatta*). *Fertil Steril*, **28**, 981–5.

Morgia F, Torti M, Montigiani M *et al.* (2006). Use of a medium buffered with N-hydroxyethylpiperazine-*N*-ethanesulfonate (HEPES) in intracytoplasmic sperm injection procedures is detrimental to the outcome of *in-vitro* fertilization. *Fertil Steril*, **85**, 1415–19.

Olds D, VanDemark NL (1957). Composition of luminal fluids in bovine female genitalia. *Fertil Steril*, **8**, 345–54.

Palasz AT, Brena PB, de la Fuente J *et al.* (2008). The effect of different zwitterionic buffers and PBS used for out of incubator procedures during standard *in vitro* embryo production on development, morphology and gene expression of bovine embryos. *Theriogenology*, **70**(9), 1461–70.

Phillips KP, Baltz JM (1996). An intracellular pH increase does not accompany egg activation in the mouse. *Mol Reprod Dev*, **45**, 52–60.

Phillips KP, Leveille MC, Claman P, Baltz JM (2000). Intracellular pH regulation in human preimplantation embryos. *Hum Reprod*, **15**, 896–904.

Quinn P, Cooke S (2004). Equivalency of culture media for human *in vitro* fertilization formulated to have the same pH under an atmosphere containing 5% or 6% carbon dioxide. *Fertil Steril*, **81**, 1502–6.

Ross M (2004). pH electrode performance. *Sensorex Tech Bull.* http://www.sensorex.com/support/more/ph_electrode_performance.

Sedlis A, Kandemir E, Stone ML (1967). Intrauterine pH of women using stainless steel contraceptive device. *Obstet Gynecol*, **30**, 114–17.

Shalgi R, Kraicer PF, Soferman N (1972). Human follicular fluid. *J Reprod Fertil*, **31**, 515–16.

Swain JE, Pool TB (2009). New pH-buffering system for media utilized during gamete and embryo manipulations for assisted reproduction. *RBM Online*, **18**(6), 799–810.

Swain JE (2010). Optimizing the culture environment in the IVF laboratory: impact of pH and buffer capacity on gamete and embryo quality. *RBM Online*, **21**(1), 6–16.

Swain JE (2012). Is there an optimum pH for culture media used in clinical IVF? *Hum Reprod Update*, **18**(3), 333–9.

Swain JE (2013). Comparison of three pH measuring devices within the IVF laboratory. *Fertil Steril*, **100**(3), S251.

ThermoScientific pH electrode handbook. http://www.webers.de/fileadmin/user_upload/dokumente/orion/R-PHREFBK-E_0109_RevA_WEB.pdf.

Vishwakarma P (1962). The pH and bicarbonate-ion content of the oviduct and uterine fluids. *Fertil Steril*, **13**, 481–5.

Will MA, Clark NA, Swain JE (2011). Biological pH buffers in IVF: help or hindrance to success? *J Assist Reprod Genet*, **28**, 711–24.

Will M, Swain JE (2012). Reducing concentration and combining zwitterionic buffers in IVF handling media allows for optimization of pH buffering capacity and supports mouse blastocyst development. *Fertil Steril*, **97**(35), 29.

Yedwab GA, Paz G, Homonnai TZ, David MP, Kraicer PF (1976). The temperature, pH and partial pressure of oxygen in the cervix and uterus of women and uterus of rats during the cycle. *Fertil Steril*, **27**, 304–9.

Zhao Y, Baltz JM (1996). Bicarbonate/chloride exchange and intracellular pH throughout preimplantation mouse embryo development. *Am J Physiol*, **271**, C1512–20.

Zhao Y, Chauvet PJ, Alper SL, Baltz JM (1995). Expression and function of bicarbonate/chloride exchangers in the preimplantation mouse embryo. *J Biol Chemi*, **270**, 24428–34.

Zander-Fox DL, Mitchell M, Thompson JG, Lane M (2010). Alterations in mouse embryo intracellular pH by DMSO during culture impair implantation and fetal growth. *RBM Online*, **21**(2), 219–29.

Zigler J, Lepe-Zuniga J, Vistica B, Gery I (1985). Analysis of the cytotoxic effects of light-exposed Hepes-containing culture medium. *In Vitro Cell Dev Biol – Plant*, **21**, 282–7.

Chapter 6

Temperature control in the IVF laboratory

Introduction

Control of temperature fluctuation is essential to successful IVF culture, from the moment of follicle aspiration, throughout all oocyte and embryo manipulations and during embryo transfer. Human microtubules are highly thermosensitive: a slight reduction in temperature can cause spindles to temporarily disassemble, and lower temperatures can cause irreversible changes to the spindle that may result in abnormal chromosome distribution when the spindles attempt to reassemble (Jurisicova *et al.*, 1996; Sun *et al.*, 2004; Wang *et al.*, 2001). Temperatures above 37 °C may induce the formation of heat shock proteins, with detrimental effects on development (Edwards & Hansen, 1996). The amount of heat loss that oocytes/embryos can tolerate without jeopardy is uncertain (Higdon *et al.*, 2008; Hong *et al.*, 2010), but a downturn in results can frequently be traced back to problems with temperature control that allow inadvertent exposure of culture systems to heat loss or overheating (Munné *et al.*, 1997; Liebermann *et al.*, 2001).

The optimal temperature range for human IVF culture is assumed to be between the limits of 36.7 °C and 37.0 °C; deviations above and below this range should be regarded as critical/dangerous. The effects of suboptimal temperature control can be manifested as fragmentation, delayed cell division, or reduced pregnancy rates due to induced chromosomal abnormalities that jeopardise implantation or result in early pregnancy loss, particularly at the pre-clinical stage (biochemical pregnancies/blighted ova).

Follicular fluid temperature

Sherbahn (2010) used thermocouple sensors to record follicular fluid temperatures at the time of oocyte retrieval, and then retrospectively analysed the clinical results after splitting the cases into three groups according to temperature:

- 36.4 – 36.9 °C = "ideal" temperature
- <36.4 °C = "low"
- >36.9 °C = "high"

The analysis clearly demonstrated that even minor fluctuations in temperature can have a significant effect on blastocyst development, implantation, and live birth rates (Table 6.1).

Evidence from porcine and bovine systems suggests that a temperature slightly lower than core body temperature (approx. 1.5 °C cooler than body temperature) may be optimal for the function of gametes and the early embryo (Leese *et al.*, 2008). Human sperm develop at a temperature a few degrees lower than body temperature, and the temperature within the human ovarian follicle has been measured as up to 2.3 °C lower than body temperature

Table 6.1 The effect of temperature fluctuation on outcome of IVF (adapted from Sherbahn, 2010, with permission)

	Low temperature <36.4 °C		Ideal temperature 36.4 – 36.9 °C	High temperature >36.9 °C	
Blastocyst development rate (from 2PN)	1408/4165 33.8%	$p < 0.0001$ vs. ideal	2331/5684, 41.0%	139/457 30.4%	$p < 0.0001$ vs. ideal
Implantation rate	482/1406 34.3%	0.0006 vs. ideal	761/1898, 40.1%	39/170 22.9%	$p < 0.0001$ vs. ideal
Live birth rate per retrieval	293/665 44.1%	0.0006 vs. ideal	484/917, 52.8%	29/77 37.7%	$p = 0.015$ vs. ideal

(Grinsted *et al.*, 1985). Lowering temperature has a profound effect on metabolism, and this has raised the question of whether clinical IVF should be performed at a temperature lower than 37.0 °C (Leese *et al.*, 2008). However, until concrete evidence emerges for human IVF, an incubation temperature of 37.0 °C is still currently preferred.

Semen assessment and temperature

Best practice guidelines for diagnostic semen analysis (Association of Biomedical Andrologists, 2012; WHO, 2010) emphasize that semen motility should be assessed at 37.0 °C as part of routine andrology work-up. Recommendations include pre-warming motility slides to prevent temperature loss via conduction, and maintaining the sample at 37.0 °C via a heated stage on the andrology microscope.

Computer-assisted sperm assessment (CASA) instruments can potentially reduce the time taken to assess motility by providing a rapid and automated assessment of sperm motion parameters, and this technology is increasingly used in andrology laboratories. Although CASA has advantages, full validation must be carried out to ensure that CASA performance provides at least the same level of accuracy as manual motility scores. Sperm are highly sensitive to temperature; some CASA systems may recommend sample assessment at room temperature, and provide an algorithm to predict motility at 37.0 °C: this type of interpretation must be approached with caution.

Principles of thermal energy

In order to ensure stable control of temperature in the culture system, it is important to be aware of potential heat loss situations and how they are affected by manipulations and environmental conditions. Heat loss and thermal equilibration are governed by basic physico-chemical principles of heat conduction and its associated parameters.

Box 6.1 Case report

Patients who produce semen samples off-site should be made aware of the importance of temperature, and instructed to maintain the sample as close to body temperature as possible during transit, delivering the sample to the laboratory as soon as possible. Below are two extreme examples that highlight the risks of off-site semen sample production:

1. A patient chose to produce his sample the night before, and put the sample in the kitchen refrigerator to "keep it fresh."
2. Another patient chose to keep his sample warm with the help of a microwave oven.

Needless to say, sperm survival was adversely affected in both instances. On-site sample production is always preferable if possible.

Conduction is the direct transfer of kinetic energy by microscopic diffusion: a temperature gradient within a substance causes particles to collide with each other. When an object is at a different temperature from another object or its surroundings, heat flows from the higher temperature region to the region at lower temperature, so that both reach the same temperature (thermal equilibrium, as described by the Second Law of Thermodynamics). Heated stages and desktop mini-incubators transfer heat to culture vessels by direct heat transfer, i.e. conduction. Figure 6.1a illustrates heat conduction from an electric cooker hob plate.

Convection is the concerted, collective movement of groups of molecules within fluids or gases – heat is carried along with the molecular flow. Convection cannot take place in solids, because there is no significant movement or diffusion of molecules within them. Convection takes place on a large scale in the atmosphere; a hot boundary layer of lower density sends plumes of hot material upwards, and cold material from the top moves downwards. Figure 6.1b illustrates heat convection from a lighted oil burner.

Large incubators usually have a water or air jacket that heats the chamber wall by conduction, and the interior provides heat by convection. Warm metal shelves then transfer heat to culture dishes by conduction.

Radiation is the transfer of thermal energy from photons in electromagnetic waves. i.e. light waves (see Chapter 7). A light source such as the light from a microscope radiates heat, and this can affect the temperature directly above the aperture in a microscope stage. This is particularly relevant during micromanipulation of gametes or embryos on a microscope heated stage that has an aperture directly above the magnification objective: the temperature over this "space" cannot be regulated by the heated stage controller.

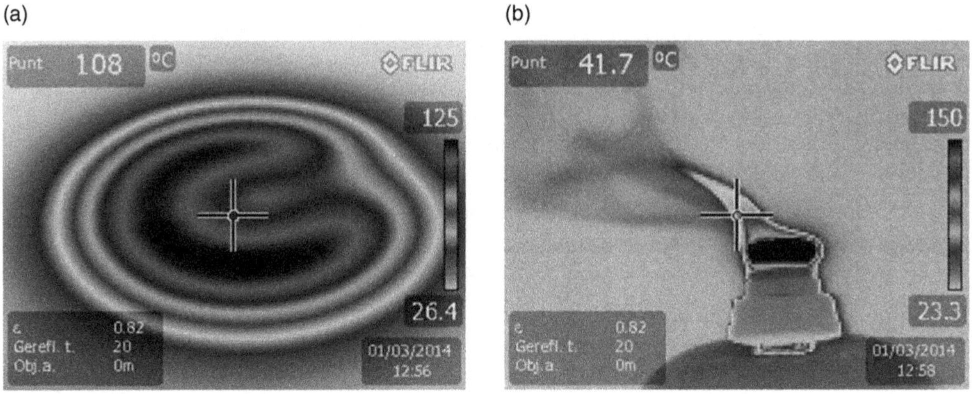

Fig. 6.1 Photo-thermographs that illustrate heat conduction and convection. (a) Heat conduction from an electric cooker hob plate; (b) Heat convection from an oil burner. *(Quartec GmbH.)* A black and white version will appear in some formats. For the color version, please refer to the plate section.

The French mathematician and physicist Joseph Fourier (1768–1830) postulated that when a temperature gradient exists, the rate of heat transfer is proportional to the temperature gradient that is present, i.e. heat flows down a temperature gradient according to the equation:

$$\Delta Q/\Delta t = hA(T_{\text{env}} - T(t)) = -hA\Delta T(t), \text{where}$$

Q = thermal energy in joules
h = heat transfer coefficient (assumed independent of T here)
A = surface area of the heat being transferred (m^2)
T = temperature of the object's surface and interior (these are the same in this approximation)
T_{env} = temperature of the environment, i.e. suitably far from the surface
$\Delta T(t) = T(t) - T_{\text{env}}$ is the time-dependent thermal gradient between environment and object.

Isaac Newton (1643–1727) translated this differential equation into his Law of Cooling, which describes the rate of change of the temperature of an object as proportional to the difference between its own temperature and the temperature of its surroundings, allowing the record of temperature/heat loss over time to be tracked:

$$T_2 = T_0 + (T_1 - T_0)\, e^{(-k^* \Delta t)}$$

T_2: Final temperature
T_1: Initial temperature
T_0: Constant temperature of the surroundings
Δt: Time difference between T_2 and T_1
k: Constant to be found

This equation can be used to estimate timing of heat loss from an embryo under different culture conditions, e.g. when a culture dish at 37.0 °C is exposed to a lower room temperature. The amount of heat loss will depend upon multiple factors, including: the air temperature, air flow, the type of culture system, the material from which the dish is made, whether a lid is placed on the dish, the base of the dish (plastic or glass), the

Box 6.2 **Practical application of Newton's Law of Cooling in forensic science**

The Law of Cooling is used by forensic scientists in order to calculate time of death:

A corpse with body temperature = 32 °C is discovered in a room.
Temperature of the room is kept constant at 20 °C.
Three hours later the temperature of the corpse dropped to 27 °C.
The observed temperatures of the corpse can be used to find "k":

$\Delta t = 3$ (hrs)
$T_0 = 20$ (°C)
$T_1 = 32$ (°C)
$T_2 = 27$ (°C)
$k = (-1/\Delta t) \ln ((T_2 - T_0)/(T_1 - T_0)) = (-1/3) \ln ((27-20)/(32-20)) = 0.1797$

Assuming the temperature of a corpse at time of death = 37 °C, and using

$k = 0.1797$
$T_0 = 20$ (°C)
$T_1 = 37$ (°C)
$T_2 = 32$ (°C)

$\Delta t = (-1/k) \ln ((T_2 - T_0)/(T_1 - T_0)) = (-1/0.1797) \ln ((32-20)/(37-20)) = 1.94$
Time of death = 1.94 hours before the corpse was discovered.

temperature of the surface on which the dish is placed, whether there is a nearby light source radiating heat (e.g. microscope), and the volume of media/oil overlay. Holding a dish in a thermal plate during transport from the incubator (standard technique in some embryology laboratories) will also significantly affect rate of heat loss. If a thermal plate is not used, the temperature of the embryologist's hand is relevant, whether or not he/she is wearing gloves, as well as how the dish is held during transport (supported under the base, or held by the sides).

The principles of conduction, convection, and radiation can be applied in order to prevent situations of heat loss, e.g. Fig. 6.2 demonstrates an experiment carried out to investigate the rate of heat loss over a 150 second time period under different holding conditions. Three 30 mm Petri dishes were filled with 2.5 ml of oil, covered with a lid, and equilibrated to 37 °C. The dishes were then removed from the incubator, and a small temperature probe inserted through a 0.5 mm hole in the lid. Readings were taken in three different situations:

1. Dish held with fingers, no contact with palm.
2. Dish left on the bench.
3. Dish held on the palm of the hand: contact with body temperature prevents significant heat loss.

Using these same principles, heat conduction properties of metal can be applied to *in vitro* culture by using a thermal coin to facilitate dish warming: the metal coin provides direct contact between its heated surface and the bottom of a culture dish (Fig. 6.3). Direct contact with a warm thermal coin helps to prevent temperature losses, and also facilitates more rapid recovery from temperature loss for the culture media in the dish.

Thermographs that illustrate the effect on thermal equilibrium in 30 mm Petri dishes are shown in Fig. 6.4. The intercalation of the thermal coin allows heat exchange by conduction instead of convection. The dishes reach the same final temperature, but there is a gain of 3 minutes with the use of a thermal coin. This experiment illustrates that the

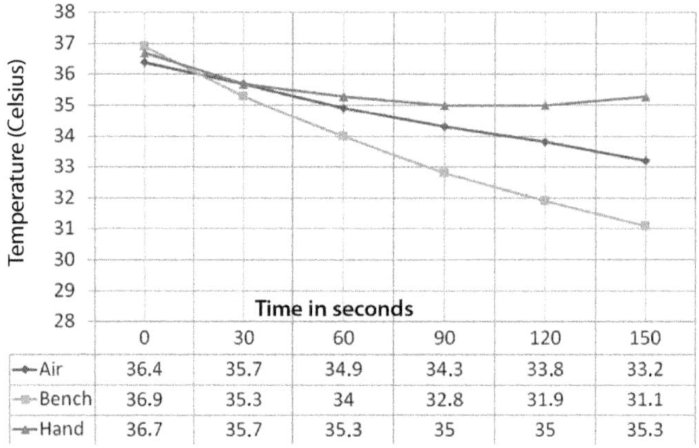

	0	30	60	90	120	150
Air	36.4	35.7	34.9	34.3	33.8	33.2
Bench	36.9	35.3	34	32.8	31.9	31.1
Hand	36.7	35.7	35.3	35	35	35.3

Fig. 6.2 Graph depicting the effect on cooling rate when a 30 mm Petri dish is taken out of an incubator at 37 °C and held at room temperature in three different ways: 1. Dish held with fingers, no contact with palm; 2. Dish left on the bench; 3. Dish held on the palm of the hand: contact with body temperature prevents significant heat loss. *(Quartec GmbH.)*

Fig. 6.3 Photograph of a 30 mm Petri dish placed on a bench (left), and with a thermal coin placed underneath (right). *(Quartec GmbH.)*

(a) (b)

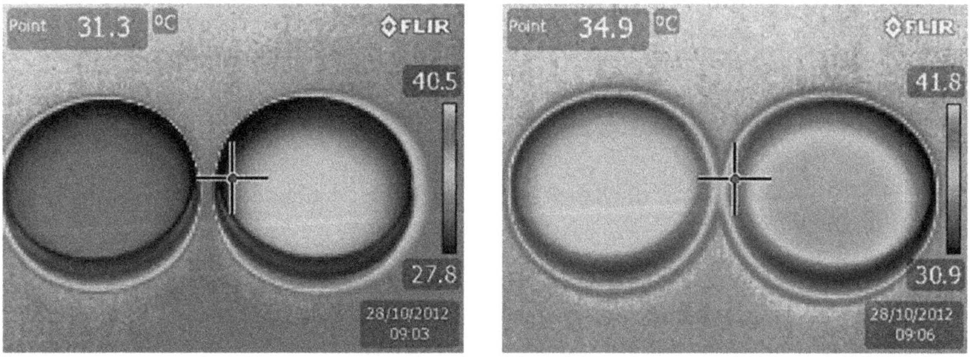

Fig. 6.4 Two Petri dishes filled with oil at the same temperature were placed on a heating stage set to 37 °C. A preheated Thermocoin (LivingCell) was placed under the dish on the right, and thermographs taken at 3-minute intervals. After 3 minutes (9.03) the dish on the left (a) had lost heat compared with the dish on the right (b). It took a further 3 minutes (9.06) for the dish without Thermocoin to reach the same temperature as the dish with the coin. The blue circles at the perimeter of both dishes indicate a lower temperature – a factor to be considered when positioning drops in an ICSI dish. *(Quartec GmbH.)* A black and white version of this figure will appear in some formats. For the color version, please refer to the plate section.

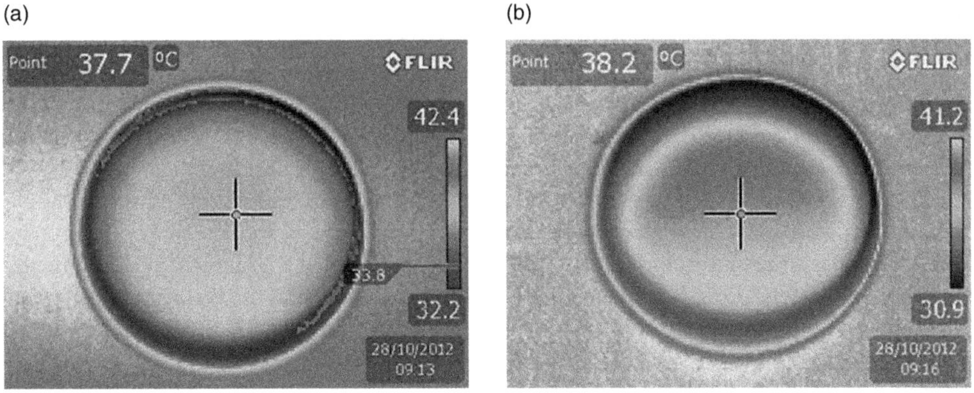

Fig. 6.5 Thermographs taken of the same dish, to illustrate artifacts that may influence interpretation of results. (a) Photo taken at a correct angle. (b) Photo taken with the camera positioned incorrectly; the angle of incidence creates an artifact, so that the upper part of the dish appears to be hotter, when it is not. A similar artifact can be created when light is reflected from a surface that is being measured. *(Quartec GmbH.)* A black and white version of this figure will appear in some formats. For the color version, please refer to the plate section.

small rim on the bottom of the dishes allows a layer of air to remain between the heating stage and the dish. A thermal coin placed under the dish allows heat to be conducted directly from the heated metal to the dish. The blue line at the border of the dishes indicates that this area is cooler: media droplets placed around the rim of the dish will have a lower temperature.

Thermographs must be interpreted with caution, as the colors perceived are subject to artifacts due to reflection of light or incorrect imaging angle: green is not "a priori" good, as demonstrated in Fig. 6.5. Convection due to fluid or gas movements can also cause unequal heat distribution.

Incubators

A reliable incubator is a crucial component of any IVF laboratory. An optimal culture system is dependent upon an environment that is stable and correct with respect to temperature, air quality, media pH, and osmolality. The IVF incubator has an influence on all of these parameters, and is therefore an integral part of the culture system. All incubators should be protected from power failure via connection to a back-up battery unit and/or generator-protected electrical outlet. Effective incubator management is a crucial cornerstone of quality management, and understanding incubator components and function is paramount in troubleshooting. Modern incubator models, their selection and management have recently been reviewed in great detail (Swain, 2014; Kelly & Cho, 2014), and only salient features relevant to troubleshooting are covered here.

Incubator models specifically designed for IVF culture have a wide range of capacities and features that use different technologies to regulate their internal environment. "Standard" large box-type incubators can accommodate a large number of cases, and independent temperature and gas probes can be placed inside them (see Fig. 8.2, Chapter 8). Many have special features, including HEPA and VOC filters, sterilization cycles, data-logging to monitor temperature, and door opening, as well as new technologies such as time-lapse imaging. Their heating system may involve direct heat (air-jacket, radiant-walled) or air convection (water-jacketed). Air-jacketed incubators recover temperature more quickly after door opening, whereas water-jacketed incubators can maintain temperature for a longer period of time after door opening or in case of a power cut. Box incubators can have the inner door divided into multi-doors (usually three or six).

Smaller benchtop-type incubators heat by direct contact (conduction), and can recover temperature and pH quickly after opening/closing. Numerous studies can be found that compare the features and clinical outcomes associated with the different types of incubators, but ultimately proper management for any type of incubator, ensuring appropriate function and optimal performance, is the most important factor in obtaining clinical success.

Some incubators are designed so that they can be "stacked", one on top of another, economising on space in small laboratories. However, this may be a "false economy": shelves that are very low or very high may be physically out of reach for some members of staff, and bending down low or reaching up high for culture dishes incurs the risk that they will be dropped. Incubators that are positioned at benchtop height avoids this risk; automatic bench height controls with dual motors can be fitted that adjust incubator and working height to the individual person: http://www.mynewlab.com/products/laboratory_desks.

Self-contained independent modular airtight chambers (MIC-101, Billups-Rothenberg, Inc., Del Mar, California) are useful as a means of customizing gas mixtures and culture conditions; they can be alcohol sterilized, and can also be convenient for isolating dishes from patients seropositive for blood-borne viruses (see Chapter 9). The chambers are flushed with the appropriate gas mixture and maintained at 37 °C in a routine laboratory incubator or warm room. Each unit has an independent culture environment of desired gas mixture and humidity, without the need for a specialised/gassed "IVF incubator." Gas ports can be modified to accommodate sensors, and with correct handling, the chambers are less susceptible to fluctuations in temperature and gas than are routine IVF incubators. The chambers can be stacked and interlocked to conserve space and provide stability, and can be a useful back-up as part of a crisis contingency plan (see Chapter 12).

Humidity

"Open" culture, without oil overlay, requires a humidified incubator atmosphere in order to prevent evaporation. This is provided by passive evaporation of water from a reservoir or pan at the bottom of the incubator. Humidity should be maintained at 95–98%. Humidity is not required for oil-overlay culture, provided that the oil has been first appropriately equilibrated (see Chapter 3).

Managing temperature control

The LED display temperature on an incubator, heating block, or heated stage must never be accepted as the actual temperature in a dish or tube. With temperature demand set at 37.0 ° C, the thermal mass of media in a culture dish can be considerably lower (Yeung *et al.*, 2004). Establishing an instrument's LED display temperature that will maintain dishes/ fluids at their optimal temperature requires careful calibration, using the appropriate independent device/system to record temperature.

Temperature mapping

A temperature gradient can often be found inside incubators, and the temperature on an upper shelf can differ by 2 °C from that on a lower shelf – the ideal location must be found by recording temperature at several different locations (Higdon *et al.*, 2007; Stoddart *et al.*, 2003; Anifandis, 2013; Walker *et al.*, 2013). All incubators (both box-type and benchtop) should be temperature mapped over a 24-hour period to determine the temperature distribution and confirm that there are no changes with time and under different conditions of use. Constant temperature mapping can be maintained, or mapping can be repeated as part of an annual validation check. Temperature mapping at different times of the year is good practice, in order to cover seasonal variations. Mapping can be carried out by placing thermocouples on every shelf of a box-type incubator, or in each compartment of a benchtop incubator in order to identify:

- hotter or cooler areas within the incubator
- variation in a single location over a period of time
- impact of regular daily interventions (e.g. opening the incubator door).

The results can be stored in portable USB temperature loggers so that the data can be downloaded. The appropriate number of probes used and their locations depend upon the way that the incubator is used. The front and back of the shelf/compartment should be measured, as well as the sites of droplets within each dish type. Figure 6.6 illustrates temperature mapping in order to validate overnight incubator function (a), and as a troubleshooting exercise during oocyte collection into a transport incubator (b).

The degree of acceptable variation must be considered in the data analysis; if necessary, a decision can be made to avoid using some areas of the incubator, labeling these shelves or compartments clearly as "not for use". Certain models of both box-type and benchtop incubator have been demonstrated to have unacceptable variation in temperature: the manufacturer should be contacted if this is found to be the case. If the problem cannot be corrected, the incubator may need to be decommissioned. Temperature mapping is always advisable when a new incubator is installed, so that any necessary corrective action can be carried out without incurring additional costs.

Temperature mapping may be affected by the incubator's location within the laboratory, giving results that may differ from those provided by the manufacturer. Air flow within the laboratory should be considered when deciding on the location of an incubator, including air conditioning units, HEPA-filter devices, door openings, etc. (NB: overheating can be just as deleterious to results as underheating – possibly even more so).

Temperature mapping of refrigerators and heated stages/working areas is also good practice. Any variations from the recommended storage temperature in media refrigerators may impact their effectiveness when used at 37.0 °C. Significant variations on microscope heated stages and heated surfaces of laminar flow hoods can be revealed by placing a dish with a thermocouple on different heated areas: the temperature within heated areas in flow hoods is notoriously inconsistent. Microscope stages with an open aperture above the light source may sometimes maintain temperature better than transparent stages that cover the aperture; this can only be revealed by rigorous and considered temperature mapping.

Door opening in large incubators causes immediate heat loss, and should always be kept to a minimum: smaller internal multi-doors are an advantage. Internal doors can be further divided into mini-doors, which allow access to one area of an incubator only, without

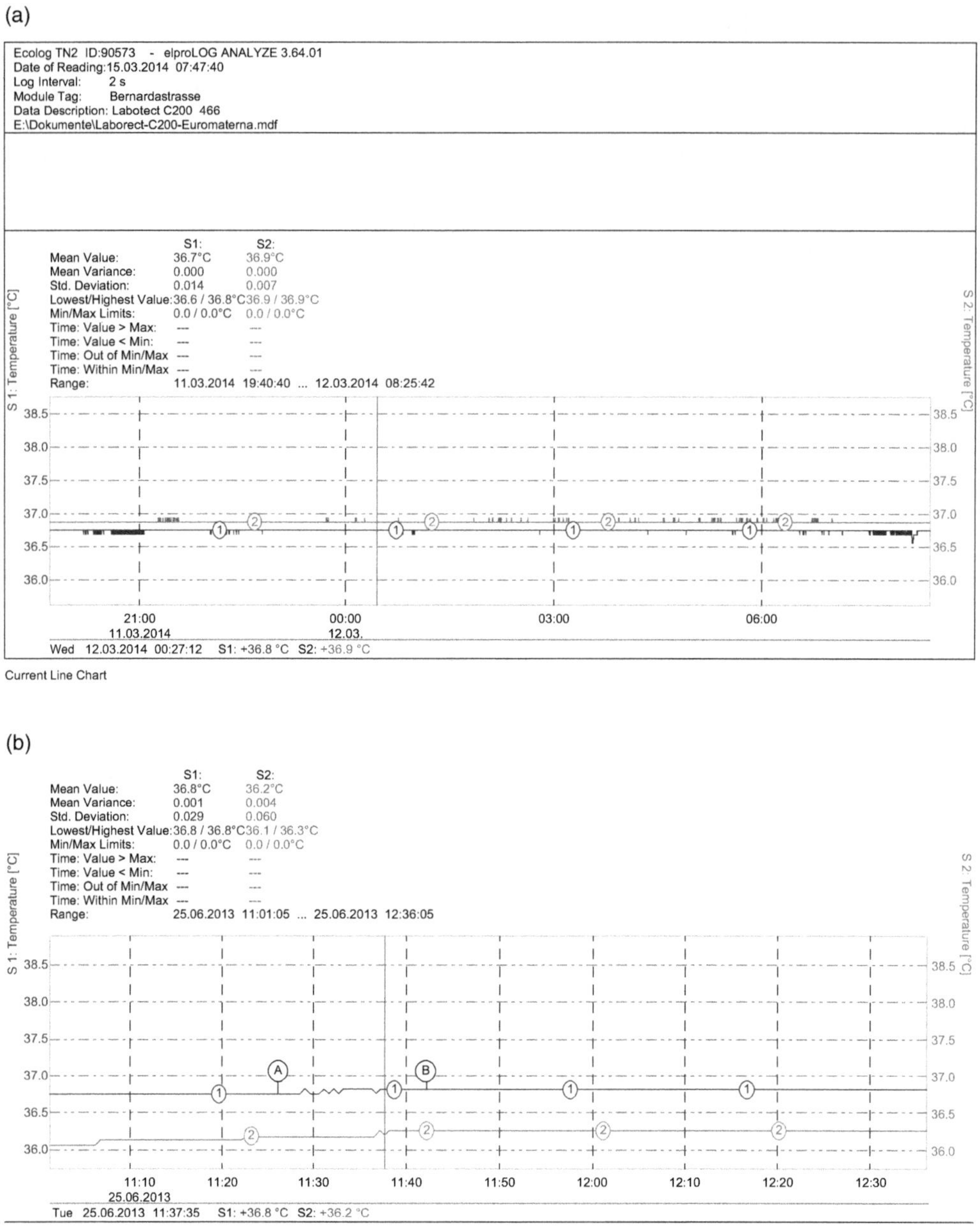

Fig. 6.6 Incubator temperature mapping.
(a) Temperature survey to validate correct overnight incubator function.
PT100 temperature probes were placed on the lower (1) and upper (2) shelves of a 150 l incubator set to 37 °C. A datalogger recorded temperature readings taken every 2 seconds between 8 pm and 9 am the following morning. The temperatures in the two locations differed by 0.2 °C: (1) = 36.6 °C – 36.8 °C; (2) = 36.7 °C – 36.9 °C.
(b) Survey undertaken as part of a troubleshooting exercise. PT100 temperature probes were used to test the function of a transport incubator during follicle aspiration and oocyte collection. The box was opened and closed each time that a tube containing fluid was removed from or placed in the box. **1** = temperature sensor in the metal heating block; **2** = temperature sensor in a cell culture tube with 5 ml medium taken from the 37 °C incubator and placed in the box 30 minutes before starting the oocyte collection. **A** = start, **B** = end of the follicle aspiration procedure. There is very little fluctuation in temperature in the heating block and in the tube with medium, but the temperature in the tube with medium was sub-optimal, never rising above 36.3 °C. The transport incubator was re-calibrated and revalidated in order to ensure optimal temperature control. *(Quartec GmbH.)*

Box 6.3 Case report

An IVF lab located in a new clinic in Africa had no air conditioning. The incubator over-heated (LCD registered 40 °C) and "blew". Fortunately, a second incubator (same model) had been installed, and the embryos were moved to the second incubator. Being reluctant to open the door, which would change the gas levels, staff "fanned" the outside of the overheated incubator in an attempt to bring the temperature down. Engineers were called in to try to fix the problem, but they had never seen an IVF incubator before. They did manage to fix the broken incubator. . .but as soon as they finished, the second one blew. This continued for 24 hours, with alternating incubators, and movement of embryos between the two. An air conditioning unit was installed the following day.

Box 6.4 Case report

The main refrigerator for culture media storage in a busy IVF lab had temperature logging, which was checked every morning. On one occasion, the temperature log revealed that the refrigerator temperature had reached 12 °C at 3 o'clock in the morning, with no apparent cause found. Although the refrigerator temperature had corrected itself and returned to 4 °C when staff arrived in the morning, a decision had to be made as to whether the media could be used. The lead embryologist took the decision to throw away the media and order a fresh supply. This involved a significant cost, but the risk of using media that had not been stored correctly was considered to be unacceptable.

disturbing other shelves/areas of the incubator. It is useful to label the outside of the external door and the internal doors with the patient's name, so that the location of a specific dish can be immediately recognized before opening the incubator door. Placing multiple dishes from too many patients into one incubator is not advisable, since constant opening of the incubator door causes fluctuation of temperature/gas, with potentially detrimental impact on other patients' dishes. A risk assessment should be performed for the number of dishes used per incubator, and used to determine the optimal number of patients treated on a daily basis. Freshly prepared culture dishes and tubes should not be placed on shelves so that they are adjacent to gametes or embryos in culture: introducing cooler plasticware and media will create a temperature gradient that may draw heat away from the pre-warmed dishes.

Removing dishes from incubators and placing them on a heated stage can involve a significant reduction in temperature, as the heated stage may not be able to maintain the appropriate temperature in the dish. When a dish is removed from the incubator and placed on a heated microscope stage set at 37.0 °C, the thermal mass of the media in the dish can fall to as low as 34.5 °C depending on the size of the dish used and the volumes of media/oil. The time taken to re-gain temperature can be as long as 10 minutes (Cooke *et al.*, 2002; Swain, 2014). The media temperature also varies with the type of culture system used, the volume of media, whether or not a lid is used, etc. Some types of four-well dish have a base that is raised from a heated stage by around 1.5 mm, so that there is no direct heat transfer between the two surfaces; heating the air under a dish in trying to achieve heat transfer can cause a localized temperature increase of up to 41.0 °C. Newer four-well culture dishes are now available to counter this problem, so that the base of the dish is in direct contact with the heated surface, minimizing heat loss.

Oil has a greater heat capacity compared with aqueous media, and using an oil overlay delays the response time in media, dampening the level of fluctuation. When gametes and

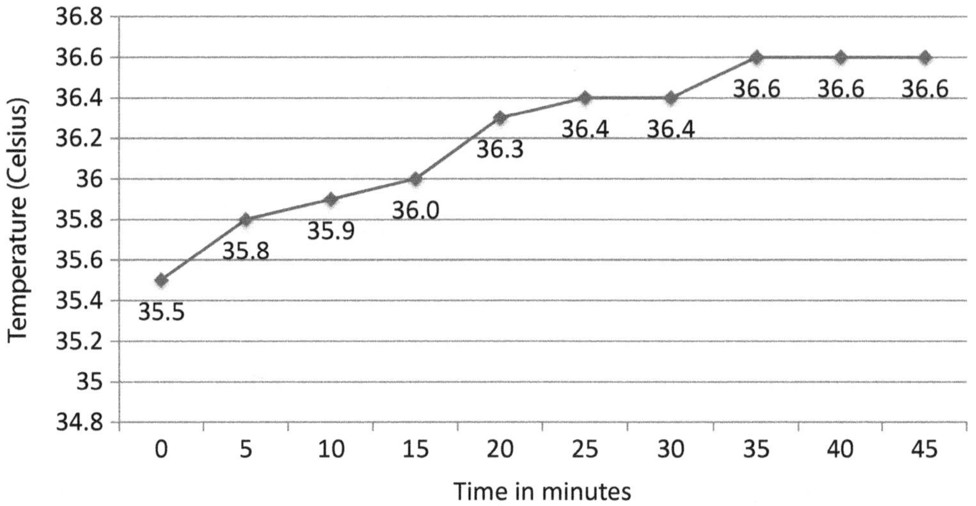

Fig 6.7 Plot of time in minutes vs. temperature in °C; a 30 mm Petri dish filled with 2 ml light paraffin oil at 34.3 °C placed in an incubator set at 37.0 °C required 55 minutes to reach a stable temperature of 36.6 °C. *(Quartec GmbH.)*

embryos are manipulated outside the incubator, an oil layer prevents the immediate effects of a decrease in outside temperature from being observed in the media layer; however, heat loss effects are noticeable after the dish is returned to the incubator. Repeated opening of an incubator door has the same effect.

Figure 6.7 plots the temperature rise in a 30 mm Petri dish with 2.5 ml oil overlay at 34.5 °C after being placed into a CO_2/O_2 incubator: a period of nearly an hour was required before a stable temperature of 36.6 °C was reached. Figure 6.8 demonstrates the effect of using a thermal coin: two dishes with oil at RT (25 °C) were placed into a 37 °C incubator, with and without a thermal coin. The thermal coin curve is 1 °C ahead of the other curve, indicating that the use of the coin decreased the time required to reach a constant temperature plateau.

If the demand temperature of the heated stage is increased in order to compensate for heat loss in a particular type of culture dish, it is important to continue using only this type/brand of dish, i.e. the stage demand temperature for a four-well dish will differ from that of round Petri dishes, with the result that samples may be overheated. Similarly the temperature of media droplets in a dish prepared for micromanipulation will differ according to whether the dish base is made from plastic or glass. The real temperature within a media droplet in a dish will be affected by:

- Distance between the base of the dish and the heated stage
- Cooling effects of air flow near the dish
- Efficiency of direct heat transfer between the heated surface and media within the dish
- Volume of media and oil overlay
- Position of the droplets within the dish (central or near the edges)
- Ambient temperature in the laboratory environment.

Droplet temperature in ICSI dishes is crucial when polarized light microscopy is used to observe the metaphase II spindle. If the droplet is not maintained at 37.0 °C, the spindle may start to disassemble (Wang *et al.*, 2001, 2002). Figure 6.9 demonstrates that the

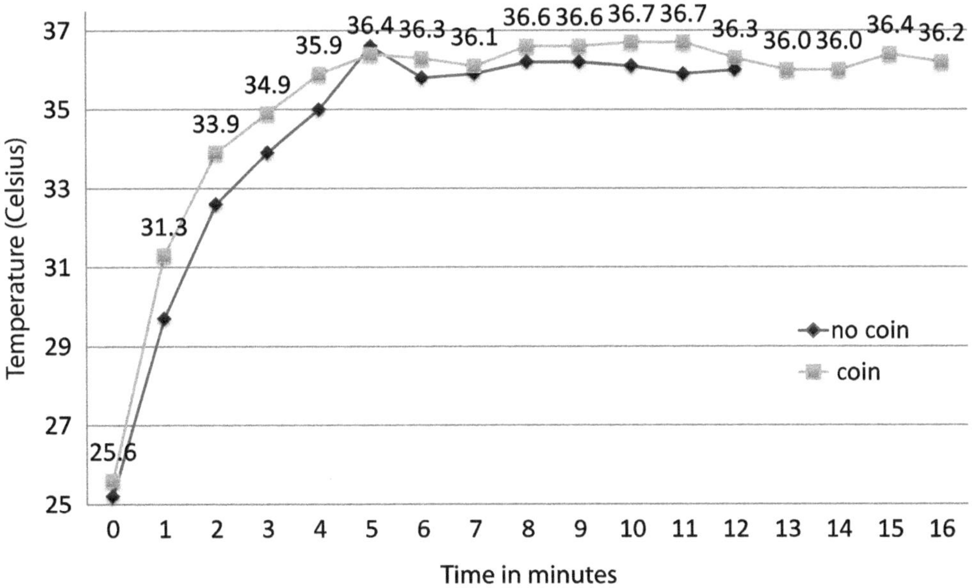

Fig. 6.8 Graph comparing the effect of using a thermal coin on time required for a 30 mm Petri dish filled with 2 ml light paraffin oil to reach 37 °C after being placed in the incubator.

temperature in an ICSI dish placed on a micromanipulation stage is not equally distributed, a factor to bear in mind when ICSI droplets are prepared.

Cooke *et al.* (2002) conducted a comprehensive series of experiments that clearly demonstrated the parameters/variables that affect temperature control in different culture systems. A digital thermocouple, calibrated and validated to an accuracy of ±0.1 °C, was used to record temperature under different experimental conditions: different types of stage warmers were assessed for their ability to maintain a constant temperature, with variations in volume of media, culture vessel lids, and type of culture incubator. Temperatures were measured under carefully controlled conditions, inside a Class II cabinet and on inverted microscope stages used for micromanipulation. Stage warmer temperatures were set according to the manufacturer's instruction books. Four different experiments were carried out:

1. Different types of stage warmers: solid glass vs. solid matte-black inside a laminar flow cabinet, and an ICSI matte-black stage warmer with a 15 mm diameter hole in the center vs. full perspex environmental chamber and heater blower with a heated glass stage ring
2. Different volumes of media/oil overlay
3. Effect of using culture vessel lid in different types of dishes
4. Small direct heat transfer incubator vs. large air convection incubator, +/-culture dish insert.

The results of these experiments are demonstrated in Fig. 6.10.

Figure 6.10a illustrates the temperature decrease in 50 μl droplets held on different microscope stage warmers, with a control reference to show temperature decrease in a dish kept on an unheated surface. The effect of media and oil overlay volumes was then compared using two different stage warmers, with the conclusion that large media volumes (1 ml) with small overlay can maintain temperature more effectively than smaller media drops (50 μl) with 1 ml oil overlay. Temperature loss was more rapid with the solid than the glass stage warmer (Fig. 6.10b).

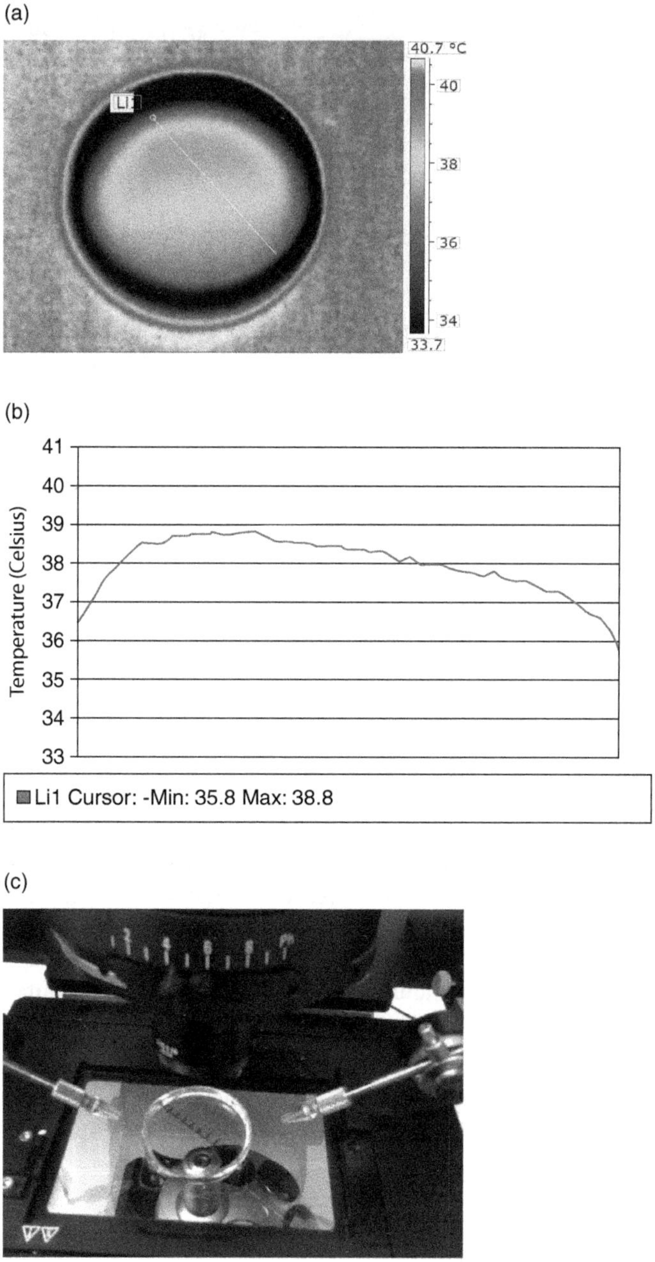

Fig. 6.9 Demonstration of unequal distribution of temperature in an ICSI dish. (a) Thermograph shows that the blue rim is colder than the center of the dish. (b) Graph plotting temperature distribution, corresponding to the white line in the thermograph (a). The shape of the curve indicates that the temperature is higher in the center of the dish than at the rim: the rim of the dish has a thicker layer of plastic, and therefore heat conduction is less efficient. (c) The red line drawn across the center of the dish corresponds to the white line in the thermograph (a), as measured by the image analyzer of the infrared camera. A black and white version of this figure will appear in some formats. For the color version, please refer to the plate section.

The stage warmer identified as being the most effective was then used to test heat loss with and without the use of a culture dish lid. The results of this experiment are illustrated in Fig. 6.10c: as expected, keeping a lid on the culture buffers temperature loss, irrespective of the volume of media used. This has implications for the use of lids when checking embryos under the microscope: small volumes of media reduce their temperature very

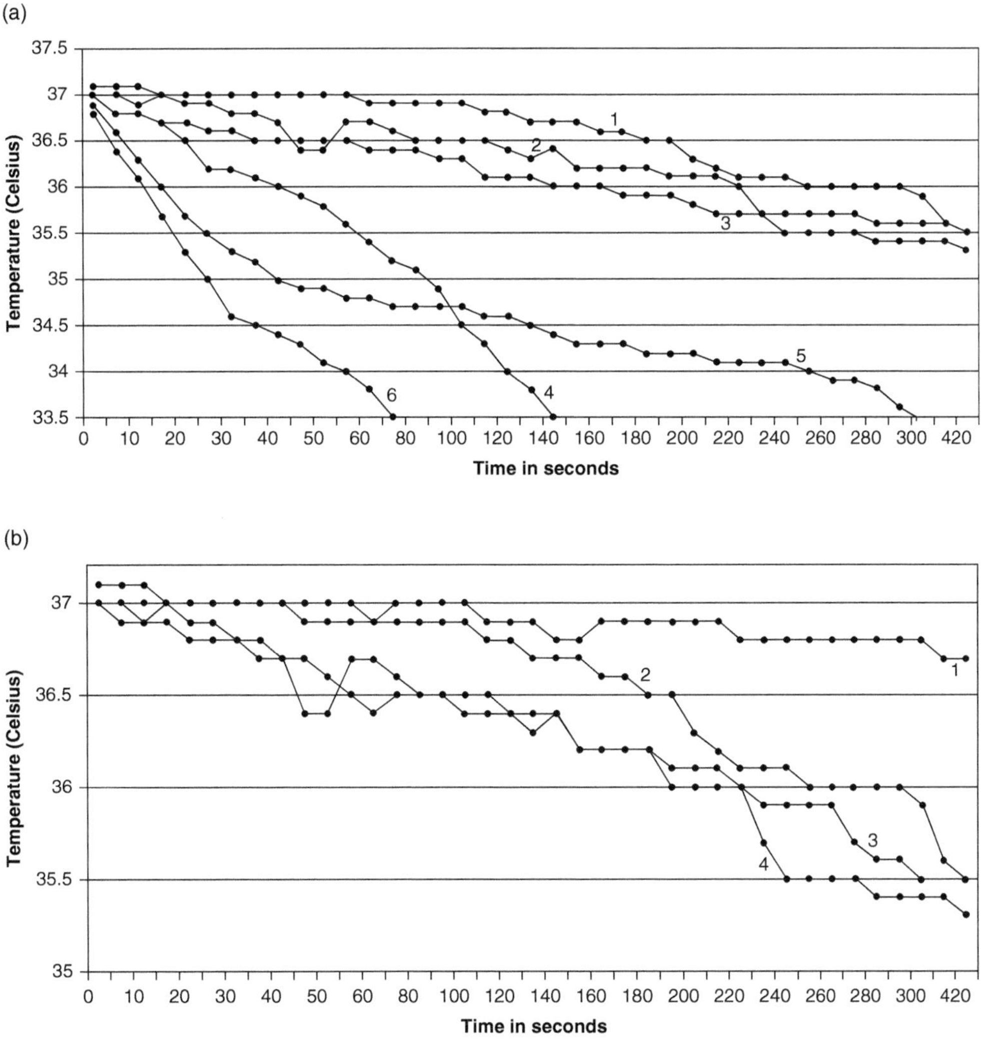

Fig. 6.10 The effect of different culture conditions and heating devices on temperature maintenance in IVF culture. *(Adapted with permission from: Cooke S, Tyler JPP, Driscoll G (2002). J Assist Reprod Genet, **19**(8), 368–75.)*

(a) Temperature decrease in 50 µl droplets on differing microscope stage warmers.

(1) Solid glass stage warmer in an airflow cabinet set at 37 °C, + lid
(2) Solid matte-black stage warmer surface in an airflow cabinet set at 37 °C, + lid
(3) Perspex environmental chamber, with heated glass stage ring set at 35.5 °C
(4) Unheated cabinet surface
(5) Matte-black microscope stage warmer with a 15 mm diameter hole removed in the center to facilitate sample viewing, set at 38 °C, with readings taken over the solid part of the stage.
(6) Matte-black microscope stage warmer set at 38 °C, readings taken over the hole in the stage

(b) Effect of culture media and oil overlay on temperature, measured on two different microscope stage warmers

(1) 1 ml of media with 100 µl oil overlay + lid, solid glass stage warmer surface set at 37 °C
(2) 50 µl of media with 1.0 ml oil overlay + lid, solid glass stage warmer surface set at 37 °C
(3) 1 ml of media with 100 µl oil overlay + lid, solid matte-black stage warmer surface set at 38 °C
(4) 50 µl of media with 1.0 ml oil overlay + lid, solid matte-black stage warmer surface set at 38 °C.

(c)

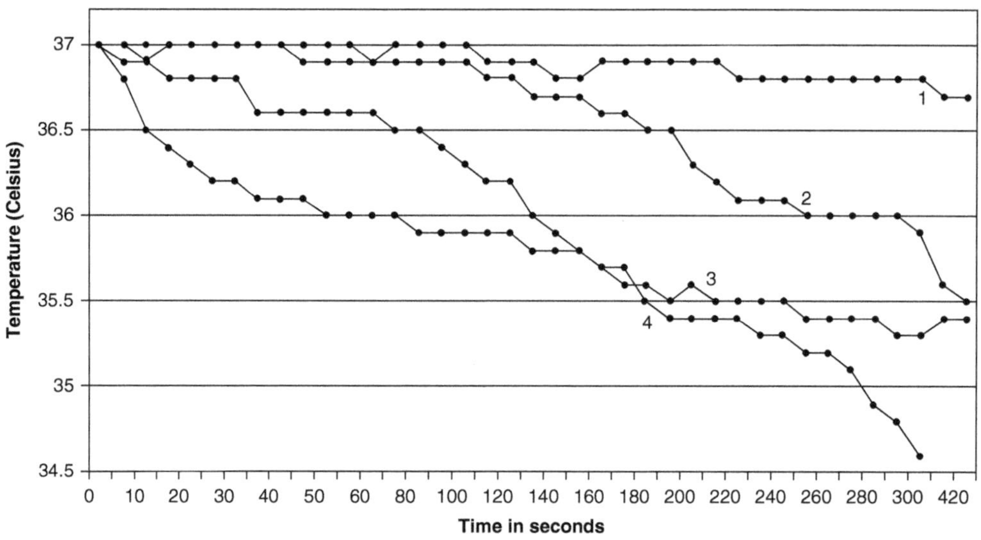

(d)

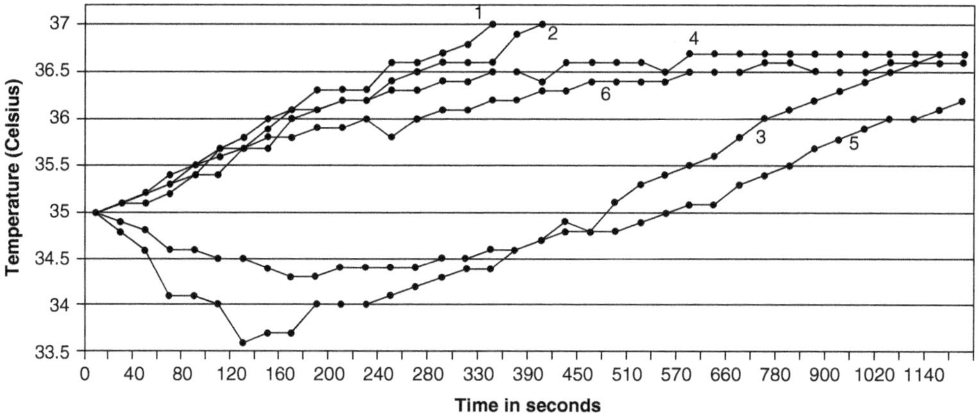

Fig. 6.10 *(cont.)*

(c) Temperature drop on a solid-glass stage warmer surface set at 37 °C with and without culture dish lid, in varying culture volumes

(1) 1 ml of media with 100 μl oil overlay + lid
(2) 50 μl of media with 1.0 ml oil overlay + lid
(3) 1 ml of media with 100 μl oil overlay without lid
(4) 50 μl of media with 1.0 ml oil overlay without lid.

(d) Temperature regain from 35.0 °C comparing a benchtop incubator with a 100 l water-jacketed incubator, with and without a recessed metal insert

(1) Benchtop incubator, 50 μl of media with 1.0 ml oil overlay + lid
(2) Benchtop incubator, 1 ml of media with 100 μl oil overlay + lid
(3) Large incubator, without metal insert, 50 μl of media with 1.0 ml oil overlay + lid
(4) Large incubator, with metal insert, 1 ml of media with 100 μl oil overlay + lid
(5) Large incubator, without metal insert, 50 μl of media with 1.0 ml oil overlay + lid
(6) Large incubator, with metal insert, 1 ml of media with 100 μl oil overlay + lid.

Box 6.5 Case report

"One of the labs that I visited regularly had a reliable and experienced team who had steadily improved their results. During one of my visits, I performed my first ICSI – and found that no oocytes had fertilized on the next day; very frustrating, as I felt it had been a good ICSI procedure. Rather than putting it down to a problem with the "sperm" (which had lost velocity during the ICSI but was still progressive), I troubleshot the whole system, and everything seemed to be fine. However... in the evening I met the clinician, who told me that he had been to a conference where he heard about heat loss during ICSI. He had entered the lab the evening before we started and turned up the heated stage temperature on the ICSI rig, without letting anyone know (there was no digital display). Thus, the oocytes had been subjected to overheating, leading to total fertilization failure. The temperature of the heated stage had to be recalibrated and revalidated, to ensure the media droplets were at 37.0 °C. The fertilization rate then returned to normal for subsequent ICSI cases."

quickly after lids are removed, whereas a dish + lid kept on an effective stage warmer for less than 2 minutes does not significantly lose its temperature.

The fourth experiment compared two different types of incubators, small direct heat transfer vs. large air convection incubators, +/− a recessed solid metal dish insert (Fig. 6.10d). Recesses in the direct heat transfer incubator kept the culture vessel surface in direct contact with the warmed incubator surface, providing effective heat transfer by conduction. Dishes placed into the large air convection incubator continued to lose temperature before starting to regain. The use of an insert acted as an intermediate method, absorbing heat from its surroundings which was transferred to the culture vessel directly.

Using an "open" culture system of 0.5 ml culture drops without overlay, Fujiwara *et al.* (2007) extended the incubator comparison to include clinical results: the experiments recorded temperature and oxygen concentration in two types of incubators (benchtop and large) following a 5-second door opening/closing. Zygotes were randomly allocated to the two systems, and subsequent cleavage and blastocyst formation rates were compared. The mini-incubator showed temperature recovery in both the chamber and the dish within approximately 5 minutes; the large incubator chamber rapidly lost temperature, and recovery time was approximately 20 minutes. Medium in the open culture dish lost temperature gradually over a 10-minute period, and recovery to 37.0 °C took approximately 30 minutes. Embryo cleavage and blastocyst formation rates were significantly higher with the use of mini-incubators. It should be emphasized that the culture systems used in the above two publications differed: although the same principles apply, each was subject to a different set of variables that would affect the end results.

Ambient temperature

Laboratory ambient room temperature (RT) is affected by heating and cooling systems; air flow patterns that change during heating and cooling cycles can create hot and cold spots in the laboratory (Bove, 2011; Butler *et al.*, 2013). Butler *et al.* (2013) used validated/calibrated wireless temperature probes (CIMScan, CIM Techniques, Inc, Beaufort SC) to evaluate the effect of ambient room temperature on equipment used in an IVF laboratory. Microscope heated stages, incubators, slide warmers, and heating blocks were monitored when the RT

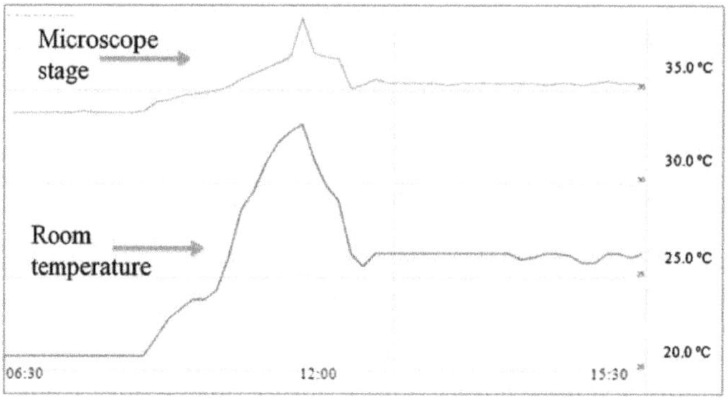

Fig. 6.11 Direct effect of a sudden upward spike in room temperature on equipment. (1) = RT, (2) = microscope stage *(Adapted with permission from: Butler et al. (2013) JARG 30(10), 1389–93.)*

Box 6.6 Case report

"I was asked to troubleshoot a lab with poor ICSI results. The micromanipulation station was at the end of a bench, below the filtered air influx and surrounded by traffic of embryologists moving around, door opening, etc. I suggested that, if the equipment could not be relocated, the micromanipulation station should be placed inside a Perspex surround to shelter the equipment from the air influx: ICSI results subsequently improved when this strategy was used."

was set to 20.0 °C, and again when RT was increased to 26.0 °C or decreased to 17.0 °C. RT was held at each temperature for 3 consecutive days, and equipment temperature was recorded at 5-minute intervals over a 12-hour period. Their results demonstrated that the temperature of all equipment, irrespective of make or model, responded unequivocally within 5 minutes to a change in RT (Fig. 6.11). Microscope heated stages and slide warmer temperatures fluctuated to a greater degree than did heating blocks and incubators. The physical location of workstations in the laboratory is important: bear in mind the spacing of air inlets and outlets, try to avoid drafts that will cause localized temperature fluctuations.

All of these experiments clearly indicate that stringent temperature control requires careful consideration, as well as accurate monitoring. Changes in RT will affect equipment, and both ambient temperature and equipment must be monitored systematically and regularly adjusted as necessary in order to maintain consistent culture conditions. The ambient temperature of an IVF laboratory may be affected by environmental and seasonal changes in outdoor temperatures: this must not be neglected when choosing the location for a laboratory.

1. Exterior walls and the top floor of a building are more likely to be affected by changes in outdoor temperature.
2. Bear in mind transfer of heat and cold through glass windows.
3. Be aware of cold and hot spots due to air flow/air conditioning/heating systems.

Measuring temperature

Heated microscope stages and heating blocks or platforms are notoriously unreliable in maintaining correct temperature control, and must be carefully calibrated and monitored,

Box 6.7 Case report

In August of 1998 the National Meteorological Institute in a European country recorded the highest outdoor temperature for 50 years, 34.4 °C. By 2pm that afternoon, the temperature in an urban IVF lab reached 32 °C, and all of the seven incubators in the lab stopped heating, activating their "low temperature" alarm: LED displays showed 36.5 °C, the lowest acceptable reading. A decision was made to place all of the manipulated gametes in the incubators, and not re-open the water-jacketed incubators (those with a double wall containing 100 l of water). All laboratory activities were stopped. Previous validation experiments after installation had confirmed that the water-jacketed incubators dropped temperature by 0.5 °C over a period of 1 hour after a power cut. The hospital's internal technical department installed a mobile air conditioning unit, but this involved opening a window in the cell culture lab that had previously been sealed. After 2 hours, RT dropped to 28.5 °C, the incubator LED continued to read 36.5 °C, and started to heat again.

During the rest of this summer, the lab was obliged to work with a window in the cell culture lab partially open – and a pigeon managed to get into the lab! The lab had to be cleaned and disinfected before work for the day could resume.

Lessons learned from this incident:

1. It is useful to know how quickly your incubators lose heat when there is a power failure, and this is not necessarily due to a power cut. These data give an approximate time window for emergency action.
2. A cell culture laboratory should be maintained at constant room temperature by appropriate air-conditioning and air filtration; opening a window automatically affects air quality inside the laboratory.
3. Installing an IVF laboratory requires more than just space – appropriate lab design is necessary.
4. Standard training of an embryologist does not include capture of a pigeon....

Fortunately, this incident did not affect clinical results; a further 2 years intervened before the institution agreed to install air conditioning in order to maintain stable room temperature....

Box 6.8 Case report

The temperature control of a water-jacketed incubator failed, and the incubator heated to 45 °C. The high-temperature alarm, which was set at 37.5 °C also failed, because the electronic circuits burned out/melted (without smoke or fire). The situation was detected when a routine morning incubator check revealed that the incubator appeared to be switched off, with no display. Fortunately, this particular incubator was used only for equilibration of oil and media, and no cultures were affected. Lessons learned from this incident:

1. Investigation revealed that no smoke or fire detectors were installed when the space was re-designed to accommodate the IVF laboratory.
2. Incubators should be monitored online, with external and separately calibrated temperature and pH/O_2/CO_2 sensors. Data must be recorded and stored.
3. The risks and consequences of incubator failure can be reduced by increasing the number of incubators, and distributing cases equally between the incubators.
4. The performance of each incubator must be recorded and compared in order to allow each incubator to be correctly validated as functioning correctly.

Box 6.9 Case report

During a time when Fertility Nurses were in short supply, an IVF clinic called upon an agency to supply nurses for weekend oocyte collections. Although the agency nurses were registered, they had no training in ART. During this period, routine morning temperature checks revealed that the heating block and the incubator in the oocyte collection room were overheating, because the set temperature had sometimes been increased. Cooling the overheated blocks takes several hours, so on each occasion a spare heating block was installed. This annoying situation continued for several months, until it was discovered that one of the agency nurses had been turning up the incubator and heating block temperatures "by hand" according to her own judgment, without mentioning or recording her actions.

Lessons to be learned from this situation:

1. Training of all personnel is extremely important, and nurses should also receive training regarding all aspects of embryology equipment that are related to their own work.
2. If possible, a dedicated and stable team is always beneficial.
3. Routine morning instrument checks are obligatory.
4. Changes in temperature set points must always be recorded.

along with ambient RT. Most importantly, optimal temperature should be maintained within the media itself (not the dish or container).

Temperatures should be recorded within the media droplet in the culture dish, under realistic working conditions: a "temperature control" culture dish can be kept on each incubator shelf for this purpose. Before taking any temperature readings, first stabilize the thermometer in a system equilibrated under the same conditions. Allow a minimum equilibration period of 1 hour before taking temperature readings for a heated stage built into a laminar flow hood, bearing in mind that compensation for cooling caused by an air stream in the hood is required. Stages heated by circulating water at 37.0 °C have been found to maintain temperature in a dish at no higher than 35 °C. Pumping water from a heated bath to a stage via tubing involves inevitable heat loss, and the use of water baths has therefore become less popular. They also introduce a risk of fungal infection from the heated water. Newer electrical heated stages are increasingly available, which avoid both of these potential risks.

Monitoring temperature effectively requires a device that has the necessary accuracy and sensitivity. Thermometers that rely on mercury or colored spirit rising in a graduated glass enclosure are very accurate, but require total immersion of the entire rising column; they can be used to measure incubator air temperature, but the temperature readings are not specific to a particular area. This type of thermometer cannot be used to measure fluctuations in small volumes – there is considerable absorption of heat when a glass bulb is submerged in a small volume. Electronic thermometers with small bulb sensors are available, but they must always be first calibrated against a reference thermometer. A disposable "Living Cell Thermometer" (LCT-01, LivingCell ApS, Denmark) is available that can give a rapid estimate of heated stage or incubator shelf temperature, with a temperature range of 35–38 °C, with increments of 0.5 °C. These may be useful as a quick visual guide, but should not be relied upon for strict temperature control.

Digital thermometers can be very sensitive (+/– 0.1 °C), and are suitable for small volumes. They consist of a processor combined with a thermocouple that measures electrical conductivity differences in the sample being tested. Three different types of temperature sensors are available:

1. Platinum Sensor PT100: very precise, covers a wide temperature range, remains stable over time, and measures in a linear range, but has the disadvantage of measuring a very small resistance (ohm) difference per centigrade. IEC 60751 (2008) defines the quality of these sensors:

   Class A sensor deviation = ± 0.15 °C at 0 °C
   Class B sensor deviation = ± 0.3 °C at 0 °C

   Therefore, laboratory temperature surveys may, in effect, be poorly monitored due to lack of precise measurement. The length of the sensor cable has an influence on the measurement.
2. NTC (negative temperature coefficient) standardized. These have high precision, react rapidly, remain stable over a long time period, and operate at a relatively high ohm resistance. They have a limited temperature range, and there is no international standardization.
3. Thermocouples or thermo-elements are composed of two different metal alloys. They are inexpensive, are recommended for higher temperatures, but are not very precise: ± 3.0 °C at 100 °C. They are susceptible to electromagnetic influences, and long probes (sensor wires) induce errors that require compensation. They are in common use, but do not have the precision to detect differences of ± 0.1 °C at *c.* 37 °C.

Thermometer calibration is essential: the accuracy of the thermometer is checked by comparing the readings with those of a standard, and any uncertainties associated with temperature measurement must be considered. Calibration is extremely important for all thermometers (total immersion and digital), and a digital thermometer must be calibrated against the thermocouple that will be used. It is very important to appreciate the difference between "calibration" and "adjustment":

- *Calibration* indicates uncertainty and deviation compared to a reference instrument.
- *Adjustment* is a technical intervention that brings the reading/display of the adjusted instrument into agreement with the reference instrument, by explicit demand.

A repeated or continued need for adjustment indicates considerable drift, and replacing the instrument is recommended.

Many countries have an Association of Testing Authorities who can test the thermometer, carry out calibration, and issue a certificate guaranteeing their work, which is accompanied by an uncertainty statement to a stated confidence level (e.g. National Institute of Standards and Technology in the US, NIST). Larger glass total immersion thermometers are valid for 10 years, and digital thermometers are valid for either 1 or 2 years before recalibration is needed. Reference temperatures must be established as correct before any readings are taken.

Practical tips for incubator settings and maintenance

(Adapted from Higdon et al., 2008, and Boone & Higdon III, 2014).

1. Follow the manufacturer's instructions for installing, setting up, and running the incubator, and make sure that you are completely familiar with all functions.

2. Always use a reliable method to calibrate the incubator – do not expect a new incubator to be "ready to use".
3. Carry out routine maintenance carefully, following manufacturers' instructions, and re-calibrate/test before using again.
4. Do not place more plasticware in the incubator than is necessary – it may off-gas (see Chapter 8).
5. Only use adhesive labels if they have been fully validated for safe use.
6. Keep a running log, with daily records of temperature, CO_2 levels, and relative humidity, noting events associated with the incubator and the surrounding environment.
7. Keep a daily temperature record of the air temperature around the incubator.
8. Try to maintain operating conditions within the incubator as tight as possible, with established maximum and minimum tolerance limits.
9. If the incubator is humidified, change water pans regularly, using sterile water.
10. Ensure regular sterilization of the incubator according to use (see Chapter 9).
11. Check periodically for mold, using blood agar plates: if mold is detected, disassemble the incubator and decontaminate.
12. Regularly change internal incubator filters, including VOC filters.
13. Always investigate if something does not seem "right" – stop using the incubator and determine whether a problem exists that can be corrected.
14. Use the incubator company's technical support team.

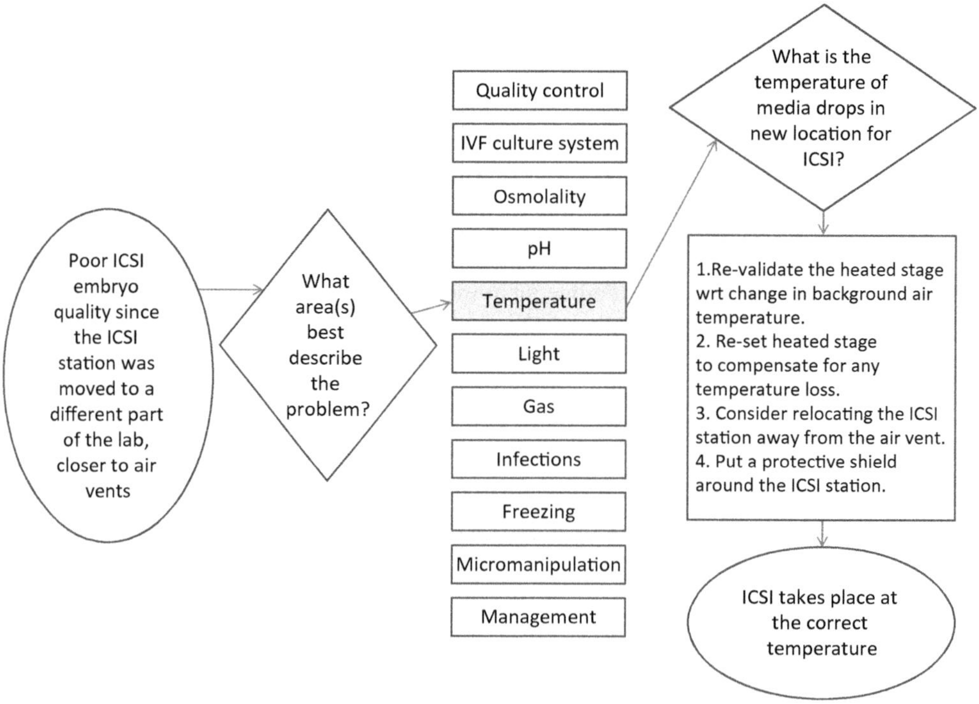

Fig. 6.12 Process map for troubleshooting temperature control.

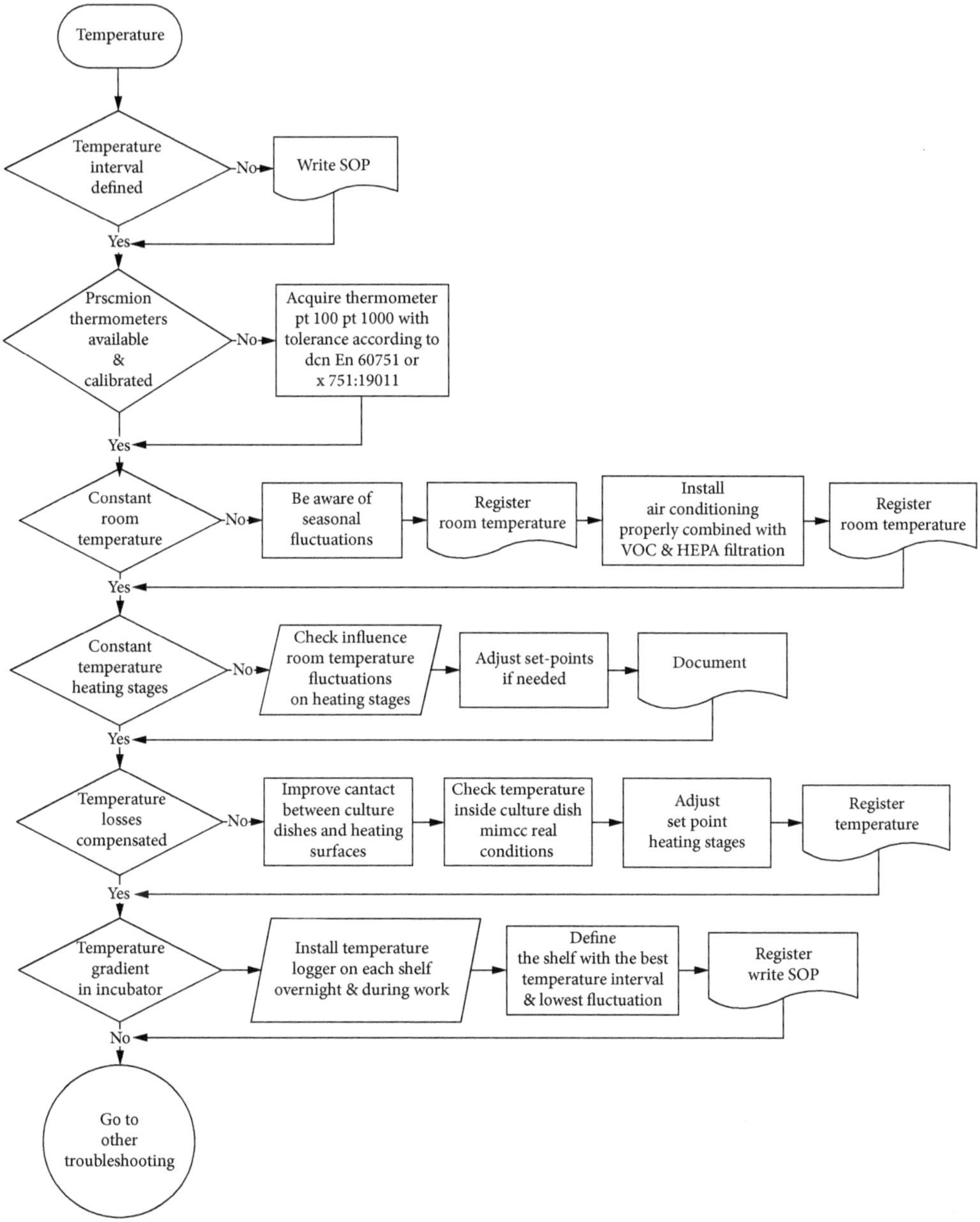

Fig. 6.13 Flowchart for troubleshooting temperature control.

Conclusion

In ART practice, there exists a Law of Three Ts: Temperature, Toxicity, and Time, all of which interact together: Some might consider that "Temperature" is the one that is often overlooked and also the most difficult to "Tame".

Further reading

Biggers JD (2001). Thoughts on embryo culture conditions. *RBM Online*, **4**, 30–8.

Bureau International des Poids et Mesures, BIPM; http://www.bipm.org/en/bipm/.

Khabani A, Tufts K, Craig L, Soules M, Scott L (2003). Cooling and warming rates in microdrops for embryo culture. *Fertil Steril*, **80**, S929–3.

Langley MT, Marek MA, Doody KM, Nackley AC, Doody KJ (2003). Importance of micro-drop temperature measurement. *EmbryoNews*, **6**, 3–5.

References

Anifandis G (2013). Temperature variations inside commercial IVF incubators. *J Assist Reprod Genet*, **30**, 1587–8.

Association of Biomedical Andrologists – Laboratory Andrology Guidelines for Good Practice Version 3 (2012). Compiled & Edited by MJ Tomlinson, SJ Harbottle, BJ Woodward & KS Lindsay on behalf of the ABA. *Hum Fertil*, **15**(4), 156–73.

Boone WR, Higdon L III (2014). Incubator management. In *Culture Media, Solutions and Systems in ART*, P Quinn (ed.), Cambridge University Press, Ch. 20, pp. 269–76.

Bove R (2011). Temperature measurement in the clinical laboratory. Good isn't good enough. *Med Lab Obs*, **43**, 36–9.

Butler JM, Johnson JE, Boone WR (2013). The heat is on: room temperature affects laboratory equipment – an observational study. *JARG*, **30**(10), 1389–93.

Conaghan J, Steel T (2008). Real-time pH profiling of IVF culture medium using an incubator device with continuous monitoring. *J Clin Embryol*, **11**, 15–16.

Cooke S, Tyler JPP, Driscoll G (2002). Objective assessments of temperature maintenance using *in vitro* culture techniques. *J Assist Reprod Genet*, **19**(8), 368–75.

Edwards JL, Hansen PJ (1996). Elevated temperature increases heat shock protein 70 synthesis in bovine two-cell embryos and compromises function of maturing oocytes. *Biol Reprod*, **55**(2), 341–6.

Fujiwara M, Takahashi K, Izuno M, *et al.* (2007). Effect of micro-environment maintenance on embryo culture after *in-vitro* fertilization: comparison of top-load mini incubator and conventional front-load incubator. *J Assist Reprod Genet*, **24**(1), 5–9.

Grinsted J, Kjer JJ, Blendstrup K, Pedersen JF (1985). Is low temperature of the follicular fluid prior to ovulation necessary for normal oocyte development? *Fertil Steril*, **43**(1), 34–9.

Higdon HL, Blackhurst DW, Boone WR (2007). Incubator management in an assisted reproductive technology laboratory. *Fertil Steril*, **89**(3), 703–10.

Hong K, Forman E, Lee H, Ferry K, Treff N, Scott R (2012). Optimizing the temperature for embyro culture in IVF: a randomized controlled trial (RCT) comparing standard culture temperature of 37 °C to the reduced more physiologic temperature of 36 °C. *Fertil Steril*, **98**, s167.

IEC 60761 (2008). Industrial Platinum Resistance Thermometers and Platinum Temperature Sensors. International Electrotechnical Commission www.global.ihs.com/IEC_Standards.

Jurisicova A, Varmuza S, Casper RF (1996). Programmed cell death and human embryo fragmentation. *Mol Hum Reprod*, **2**, 93–8.

Kelly E, Cho T (2014). Incubators old and new. In *Culture Media, Solutions and Systems in ART*, P Quinn (ed.), Cambridge University Press, Ch. 19, pp. 258–69.

Kelly PB, Deignan KC, Emerson G, Mocanu E (2009). Assessment of temperature variation during *in vitro* culturing: a comparison of different IVF micro-environments. *Fertil Steril*, **92**, S234.

Leese HJ, Baumann CG, Brison DR, McEvoy TG, Sturmey RG (2008). Metabolism of the viable mammalian embryo: quietness revisited. *Mol Hum Reprod*, **14**(12), 667–72.

Liebermann J, Graham J, Han T, Carter J, Tucker M (2001). Temperature fluctuation of microdroplet in IVF. In *Third Biennial Alpha Conference*, 8–11 Sept. 2001, New York City, New York; Abstract book, O-12.

Munné S, Magli C, Adler A, *et al.* (1997). Treatment-related chromosome abnormalities in human embryos. *Hum Reprod*, **12**, 780–4.

Sherbahn R (2010). Assessment of effect of follicular fluid temperature at egg retrieval on blastocyst development, implantation and live birth rates. *Fertil Steril*, **94**(4), S68–9.

Stoddart N, Rho H, Maisel H, Zhao T, Tarantino S, Fleming S (2003). Influence of the position of the culture dish within the incubator chamber and temperature of culture media on clinical outcome following Assisted Reproductive Technology (ART) treatment. *Fertil Steril*, **80**, s164–5.

Sun XF, Wang WH, Keefe DL (2004). Overheating is detrimental to meiotic spindles within *in vitro* matured human oocytes. *Zygote*, **12**, 65–70.

Swain JE (2014). Decisions for the IVF laboratory: comparative analysis of embryo culture incubators. *RBM Online*, **28**, 535–47.

Walker MW, Butler JM, Higdon HL, Boone WR (2013). Temperature variations within and between incubators–a prospective, observational study. *J. Assist Reprod Genet*, **30**, 1583–5.

Wang WH, Meng L, Hackett RJ, Odenbourg R, Keefe DL (2001). Limited recovery of meiotic spindles in living human oocytes after cooling–rewarming observed using polarized light microscopy. *Hum Reprod*, **16**(11), 2374–8.

Wang WH, Meng L, Hackett RJ, Oldenbourg R, Keefe MD (2002). Rigorous thermal control during intracytoplasmic sperm injection stabilizes the meiotic spindle and improves fertilization and pregnancy rates. *Fertil Steril*, **77**(6), 1274–7.

WHO (2010). *Laboratory Manual for the Examination and Processing of Human Semen*, Fifth edn. World Health Organization, Department of Reproductive Health and Research.

Yeung QSY, Briton-Jones CM, Tjer GCC, Chiu TTY, Haines C (2004). The efficacy of test tube warming devices used during oocyte retrieval for IVF. *J Assist Reprod Genet*, **21**, 355–60.

Chapter 7

The effect of light on embryos and embryo culture

Kimball O. Pomeroy and Michael Reed

Introduction

The natural incubator of the mammalian embryo, the uterus, has been fine-tuned for millions of years. Homeostatic measures are in place to provide an environment that changes little, despite an external environment that is often quite variable. IVF laboratories use modern mechanical incubators that have to some extent successfully replicated this environment (Swain, 2014).

The goal of both natural and artificial incubator systems is the same: to protect the embryos from insults that might result in consumption of the embryo's precious energy and other resources, or even in the demise of the embryo. The environmental control in incubators has been focused largely on temperature and pH. In the body, changes in osmolality, pH, and temperature are normally small and gradual, but in an IVF system these excursions can be relatively large and rapid. Perturbations *in vitro*, not normally present *in vivo* in the uterus may result in stress to the embryos. Other environmental factors must also be taken into consideration in the process of moving embryo culture into this artificial environment, including exposure to toxicants, off-gassing of volatile organic compounds (see Chapter 8), exposure to electromagnetic fields, generation of reactive oxygen species, and light exposure.

In order to work safely whilst manipulating embryos and monitoring their quality, the embryo must be exposed to light – a potential environmental factor that is not relevant during *in vivo* culture. Since human embryos essentially grow in a dark uterine environment, measures to protect them from light may not have developed during the evolution of the species. ART procedures involve exposing embryos, sperm, and oocytes to light during oocyte retrieval, placing dishes into and out of incubators, during microscopy for ICSI, fertilization checks, morphological assessment, and during embryo transfer. The use of time-lapse imaging has also increased the frequency at which developing zygotes and embryos are exposed to light, which can be as often as every 10 minutes.

Although there is evidence that some mammalian fetuses, and even the late human fetus, may be exposed to low intensities of light *in vivo* (Jacques *et al.*, 1987; Del Giudice, 2011), oocytes and embryos may not have a functional system to either repair or protect against potential light damage during IVF; it is therefore possible that irreparable harm may be done. Ethical and financial considerations of experimenting with human embryos make it unlikely that their interaction with light will ever be adequately studied in the future.

Troubleshooting and Problem-Solving in the IVF Laboratory, ed. Kay Elder, Marc Van den Berg, and Bryan Woodward. Published by Cambridge University Press.

Photo-oxidation

Several variables are important when measuring light: duration of exposure, intensity of exposure, and wavelength. Light intensity is often measured in lux, but this measurement represents the intensity as measured by the human eye, and is not suitable for non-visible wavelengths. Lux is also a poor measurement as it does not take into account the duration of light exposure. A better measure of intensity is irradiance (watts/m^2) which, as a measure of power (energy/unit time; joule/sec), also includes the duration of exposure. Unfortunately, most of the studies that investigate the influence of light on cell culture do not measure irradiance, making it difficult to determine the amount of light the cells were exposed to in these experiments.

Light might affect a cell in several different ways. There may be a direct effect where light "stresses" the cell, activates stress genes or even damages DNA directly via ionization. Light may also indirectly affect cells by oxidation of components in the media or oil, changing a neutral or even beneficial component into a toxicant. This indirect method can occur via photo-oxidation – a chemical reaction between light and components of culture medium and oil. Light has been implicated in the oxidation of oil used in the culture of human embryos (Otsuki *et al.*, 2009). The mechanisms involved in photo-oxidation of media components and oil may also be present in photo-oxidation of sperm and oocyte membranes (which are lipids, as are oils), producing changes in these membranes that could potentially inhibit fertilization or affect morphology. Light has also been shown to induce production of hydrogen peroxide (H_2O_2), a reactive oxygen species (ROS). ROS with cell toxicity are generated when media containing HEPES (N-2-hydroxyethylpiperazine-N′-2-ethanesulfonic acid) and riboflavin are exposed to light (Hill Jr. *et al.*, 1960a; Stoien & Wang, 1974; Wang & Nixon, 1978; Wang, 1975; Zigler *et al.*, 1985; Zieger *et al.*, 1991; Grzelak, 2001).

Sea urchin embryos that incubate in tidal pools are naturally exposed to light, and ultraviolet B radiation in the 290–320 nm wavelength can cause DNA damage, oxidative stress (Lesser *et al.*, 2003) and damage to proteins and membrane lipids (Halliwell, 2006). The intensity of light that can cause damage is in the range that sea urchin embryos are naturally exposed to during embryonic growth in tidal pools. When sea urchin embryos are exposed to excessive light, resulting DNA damage causes the cells to be retained in the G1/S phase of the cycle in order to repair the damaged DNA, with consequent delays in cell division and development. If the DNA damage cannot be repaired, the sea urchin cells will undergo apoptosis and cell death. Sea urchin DNA damage can occur not only due to the direct effect of light on DNA, but also indirectly via ROS (Lesser *et al.*, 2003).

Sea urchin embryos have developed systems to protect them from damage due to light exposure – for example, expression of the antioxidant superoxide dismutase. They can also increase the expression of p53 and p21 genes, both of which play a role in apoptosis or cell cycle arrest. These genes can help to either repair damage or eliminate damaged cells. p53 is also a cell checkpoint gene that allows the repair or removal of cells in response to DNA damage, hypoxia, ROS, or mutagens.

Sea urchins have developed these repair mechanisms to light damage in response to natural selection over millennia. Although humans have the same type of repair machinery and cell cycle checkpoints, it is not known whether these systems are functional and whether they can repair any potential damage from light exposure encountered during *in vitro* culture.

Light toxicity, tissue cultures, and non-primate embryos

One of the first indications that light could be toxic to cells was a report that protozoans placed in an acridine dye solution were killed by light exposure (Hill Jr. *et al.*, 1960b). This was not a direct effect of light, but was due to a modification of the chemical properties of acridine orange by light that resulted in decreased protein synthesis, inhibition of replication, and growth arrest. Further studies have shown that light can also indirectly damage mammalian cells via photo-oxidation (Stoien & Wang, 1974; Wang, 1975). One study showed that photo-oxidation of media components (riboflavin, tryptophan, and tyrosine) resulted in the production of toxic H_2O_2. The addition of HEPES to murine thymocyte culture medium RPMI 1640 markedly increased the production of cytotoxic products when the medium was exposed to visible light (Zigler *et al.*, 1985; Zieger *et al.*, 1991).

There are also several examples where light has been implicated in harming either gametes or embryos. The first observation of the direct effects of light on mammalian embryos was by Daniel (Daniel, 1964), who observed that rabbit ova exposed to light for up to 12 hours displayed retarded cleavage rates. Cleavage rates were affected even with the use of various filters: red filters seemed the most effective in ameliorating the cleavage inhibition. Others have shown (Schumacher & Fischer, 1988) that earlier stage embryos are more sensitive to light than later stages; day-1 rabbit embryos exposed to light for as little as 1 hour can slow their proliferation, as measured by decreased thymidine incorporation. It should be noted that at least one study (Bedford & Dobrenis, 1989) reported no negative effects on embryonic development or implantation in rabbit embryos exposed to 3250 lux of fluorescent light.

Hamster embryos are also quite sensitive to light. Failure of hamster embryos to progress normally after injection of sperm (ICSI) has been attributed to the toxic effect of light on hamster ova. The first birth of hamsters using ICSI was achieved only when the microscope light source was filtered, using red light in a darkened room (Yamauchi *et al.*, 2002). Research has shown that exposing hamster oocytes to fluorescent lighting for only 1 hour can inhibit normal meiotic development (Hirao & Yanagimachi, 1978). Light has been shown to damage hamster chromosomes, the most common defects being either failure of the chromosomes to develop properly after metaphase, or the formation of numerous small pronuclei. In most cases, the second polar body was not extruded after sperm injection. Wavelengths in the 470–480 nm range (blue–violet–ultraviolet) were the most harmful, and a red filter was protective. Sunlight, UV light, and fluorescent lamps were more detrimental than incandescent lamps.

It appears that, at least in the hamster, short wavelengths of light can disrupt the spindle apparatus and chromosomes. In another study, light was shown to affect the early embryonic development of hamster embryos (Umaoka *et al.*, 1992). As little as 30 minutes of exposure to light (380–760 nm) blocked the development of zygotes to the two-cell stage. Embryo development was also inhibited when two-cell and eight-cell hamster embryos were exposed to as little as 5 minutes of light.

A further study demonstrated the effect of light exposure at different intensities and wavelengths on hamster two-cell embryos (Oh *et al.*, 2007). Concentrations of toxic ROS and heat shock protein 70 (an inducible cell stress gene) were measured during assessment of early embryonic development. Development to the blastocyst stage was inhibited by exposure to 900 lux when compared to 200 lux. Wavelengths in the 445 to 500 range (visible blue) also reduced development to the blastocyst stage, increased the expression of HSP70 and resulted in higher levels of ROS compared to embryos exposed to 620–750 nm light (visible orange/red).

Light can affect the development of mouse embryos (Takenaka *et al.*, 2007) as well as hamster and rabbit embryos, although to a lesser degree. Early embryonic development was not affected by exposure of mouse embryos to cool white fluorescent light at 1200 lux for 15 minutes, but increased internal levels of ROS were detected under the same conditions. The concentration of light-induced ROS was also lower in mouse embryos compared with hamster embryos. Mouse embryos exposed to cool white light had a higher number of apoptotic cells in blastocysts than embryos shielded from light. Implantation of normal blastocysts transferred into pseudo-pregnant females was also lower in mouse embryos exposed to light compared to shielded embryos (66% vs. 42%).

Another study examined the effect of light on the *in vitro* production of bovine embryos (Korhonen *et al.*, 2009). Embryos were exposed to either filtered light (498 nm to 563 nm) or unfiltered light (halogen lamp) while being examined with a microscope. No differences were detected in embryo development or in the number of cells in blastocysts, but unfiltered-exposed bovine embryos had higher levels of the inducible stress protein, HSP70. Although light does not apparently affect early development of mouse or bovine embryos overtly, there is a toxic affect as measured by induction of ROS, HSP70, or impairment of implantation (in the mouse).

Light exposure in the human IVF laboratory

In 1998 Robert Edwards commented on his concerns regarding handling of gametes and embryos *in vitro*, and exposure of embryos to light:

> Light has also been one of my major concerns ever since IVF began. We were aware of the many papers on mammals published by embryologists on the evolution of reactive oxygen species in response to light exposure, and its deleterious effects on embryo growth. We could not afford any risks with human embryos to be replaced into the mother, so we used green filters routinely to remove some of the light radiation, lower the light intensity and produce a more acceptable colour for the eye by modifying the harsh artificial light from the microscope. The potential effects of light concerned me in another way. During transfer, gynecologists often used an intense operating theatre light to shine on the cervical os. Yet this was where the embryo is passed during the transfer process. At the last moment, after hormone stimulation, oocyte collection, fertilization and cleavage *in vitro*, these precious embryos could be exposed to an intense light which might impair their ongoing development. We therefore dimmed this light during transfer to avoid any damage to the embryos in the last stage of their ex-utero existence. Several investigators have disparaged my attitude, and they may even be right when they claim that human embryos can tolerate this degree of intense light exposure. But I have never seen any evidence on this point from these investigators, and it is surely better to be safe than sorry. So I still use many of these precautions. (Edwards, 1998, pp. 271–2, reprinted with permission)

A retrospective study of gamete intrafallopian transfer (GIFT) procedures revealed that different laparoscopic light sources (halogen vs. xenon) had a significant effect on pregnancy rates: 50% pregnancy rate in the cycles that used a halogen light source, compared with 9% using a xenon light (Evans *et al.*, 1999). The authors further confirmed that the xenon light source emitted more ultraviolet light than the conventional halogen light source, and suggested that the difference in pregnancy rates could be due to a detrimental effect of light on the gametes during the GIFT procedure.

In the context of minimizing additive or interactive stressors *in vitro*, effects of lighting may be added to temperature, pH, ionic, metabolic, and other environmental excursions

Table 7.1 Typical sources of visible light in the laboratory

Windows
Filtered, curtained, shades, blinds
Ceiling lights
Fluorescent; cold white, warm white
Incandescent; dimmable and single intensity
Lamps
Floor, desktop
Hoods/cabinets
Microscopes
Inverted
Dissecting
Time-lapse
Direct effects of light:
Mineral oil overlay
Culture medium components
Plasticware
Indirect effect of light:
ROS
Gene transcription
Culture medium breakdown products
Oocyte retrieval
Direct and indirect room lighting, hoods, other
Surgery lamps
Microscope
Manipulation and routine handling
Direct and indirect room lighting, hoods, other
Microscope
Embryo transfer
Direct and indirect room lighting, hoods, other
Microscopes
Headlamps, floor lamps, surgery lamps

that influence developmental competency. There are no well-designed studies utilizing human gametes and embryos *in vitro* to evaluate the impact, if any, of the type of lighting, duration of exposure, or exposure to specific wavelengths; the information available on this topic is derived from the previously mentioned animal model studies, where the tolerance to alterations in wavelength (nm) and intensity of light exposure varies according to developmental stage, species, and environmental temperature.

In the early years of IVF, laboratories were adapted from rooms originally tasked for other procedures, ranging from closets with no lighting to operating theaters with variable sources of light, rooms with incandescent lighting that might have variable intensity controls, and rooms with only standard fluorescent cool wavelength lights; the type, quality, and intensity of lighting was not often taken into consideration (see Table 7.1). In fact, there

Table 7.2 Sources of potentially toxic light exposure in the IVF laboratory

Type	Source	Comments
Sunlight	Windows	Includes 400 t0 500 nm
Fluorescent	Indoor lighting	440 to 500 nm via mercury arc
Compact fluorescent	Indoor lighting hoods	400 to 700 nm
Incandescent	Indoor lighting	Some light in 400 to 500 nm
Halogen	Indoor lighting Microscopes	Depends on bulb; can be coated to filter out wavelengths
LED	Microscopes Indoor lighting	Depends on LED type
Mercury arc	Microscopes	300 to 600 nm range

were few to no standards or laboratory prototypes available, nor was there a consensus as to what constituted the "ideal" laboratory environment.

With regard to human clinical IVF, there is one early publication discussing prospective changes made in the laboratory design to improve the laboratory environment; two parameters were considered, oxygen tension and lighting in the laboratory (Noda *et al.*, 1994). Although this study did not follow through to embryo transfer and gestational parameters, supernumerary surplus embryos were allowed to develop under (1) low oxygen conditions (5%) and (2) low intensity lighting (20 and 100 lux, as measured from ambient and microscope lights, respectively). The authors noted increased development to the blastocyst stage in the low oxygen tension and low light conditions compared to culture under room air and standard lighting conditions.

A landmark study (Ottosen *et al.*, 2007) provided measured, replicated baseline data for light intensity and duration of exposure to light under working clinical IVF laboratory conditions; a laboratory with windows, a laboratory without windows, use of various wavelength filters applied to microscopes, and importantly, technician preferences for microscope light intensity settings.

Embryologists should be aware that light in the 400 to 500 nm wavelength range (indigo–blue–green, see Fig. 7.1) has the greatest potential for affecting gametes and embryos. These wavelengths are found in most light sources as indicated in Table 7.2. The bottom line is that all light sources that do not include a filter probably emit some light in a range that could potentially be damaging to gametes and embryos.

Is it practical or possible to reduce the wavelengths in the critical spectra – 400 to 500 nm, without compromising patient and technician safety? Under variable, but working lighting conditions, Ottosen and colleagues demonstrated that plastic polystyrene dishes did reduce or absorb small, but not significant amounts of energies in the 400 to 500 nm spectra. Short wavelength filters applied to halogen bulb microscopes and room bulbs (background fluorescent and natural lighting) significantly reduced the light energies in this critical spectral range. Duration and intensity of light exposure were also considered important, and were modeled according to the preference of each technician. The authors concluded that reducing ambient light intensity (while maintaining light intensity comfortable for the technicians), reducing duration of light exposure to gametes and

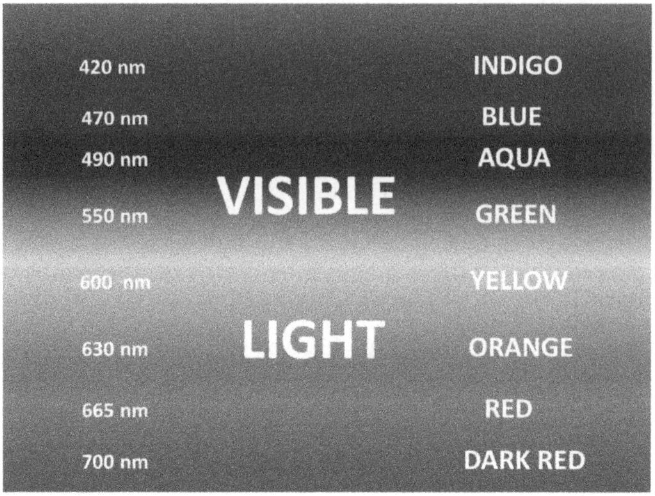

Fig. 7.1 The spectrum of visible light; numbers on the left represent wavelength in nanometers. Ultraviolet light has wavelengths between approximately 310 nm and 100 nm, with photon energies from 4 eV to 12.4 eV. Wavelengths in the 400–300 nm range are known as "near ultraviolet" (visible to birds, insects, and fish), and have photon energies of 3.10–4.13 eV. At the other end of the spectrum, infrared (IR) light wavelengths extend from the red edge of the visible spectrum at 700 nm to 1 mm, with photon energies 1.24 mEV–1.7 eV.
Most of the thermal radiation emitted by objects near room temperature is infrared radiant energy. A black and white version of this figure will appear in some formats. For a color version, please refer to the plate section.

embryos, and specifically targeting microscope lighting by adding red or amber filters would provide the greatest benefit.

Using a green bypass filter (such as the Schott VG-9) could provide a simple strategy to reduce embryo light exposure from a microscope, until the effect of light on human embryos is better understood. A green bypass filter can protect embryos from deleterious effects of both blue and near infra-red wavelengths. A green filter is more effective than red, as the human eye is more sensitive to green than to red (Parr, 2001).

From a practical standpoint, lighting in the laboratory should be sufficiently bright for personnel safety, at the same time ensuring the safety of the gametes and embryos handled in the laboratory. Blue wavelength filters create a visible spectrum perceived as amber to yellow light; in this environment, color labels can be washed out or altered, pH monitoring via phenol red in culture medium can be compromised, and the filters reduce the intensity of available lighting, making it necessary to employ additional incandescent lamps in some laboratories.

Reactive oxygen species

Potentially negative effects of light on ova and embryos can be reduced via two strategies: (1) reduce exposure to light via filters on microscopes and light fixtures to remove the blue spectrum (mentioned above) and (2) mitigate damage from light-induced ROS by adding antioxidants to culture media.

ROS are generated as an intrinsic consequence of cell metabolism, e.g. mitochondrial function, where the potentially damaging nature of these molecules is balanced by the inherent antioxidant capacity of the cells. When this balance is interrupted, an overabundance of more potent radicals (hydroxyl and H_2O_2 compounds) can damage cells, specifically targeting lipids and DNA. Gametes and embryos *in vitro* are exposed, transiently or continuously to supra-physiological oxygen, temperature and pH excursions; exposure to various light energies can be responsible for adding to the total oxidative cell load *in vitro*. Male and female reproductive cells are sensitive to increased ROS exposure *in vivo* and *in vitro*, and in some cases, reproductive potential may be impaired (Agarwal & Allamaneni, 2004; Agarwal *et al.*, 2006).

The specific targets for damage-inducing ROS compounds *in vitro* has been debated. Mammalian sperm plasma membrane lipids can be damaged following light-induced increased ROS production. Oxygen radicals were produced in significantly greater quantities following exposure of sperm membranes to blue spectra (400–505 nm) compared to red spectra (600–800 nm); the authors speculate that the mechanism involves membrane lipids directly, and not via an external catalyzed process (Lavi *et al.*, 2012). Ram sperm were exposed to white light (400–800 nm), red light (660 nm), blue light (360 nm) and UV light (294 nm) *in vitro*. Processed, light-exposed sperm were then used for *in vitro* fertilization. Motility of ram sperm was compromised by exposure to white, blue, and UV spectra, and embryo viability after IVF was highest for sperm exposed to control (darkness) and red spectra. Additionally, lower ROS concentrations were generated by exposure to the longer (red) spectral wavelengths (Zan-Bar *et al.*, 2005).

As previously mentioned, light has been linked to increased production of ROS and to DNA damage. Human embryo fragmentation (Yang *et al.*, 1998) has been linked to apoptotic processes, which may be affected *in vitro* by ROS production; furthermore, ROS production increases as oxygen tension increases. The detrimental effect of increased ROS may be manifested at the mitochondrial level; the antioxidative capacity of the culture medium and the embryo itself may counter some of these negative ROS effects (Catt & Henman, 2000; Guerin *et al.*, 2001). It is thus possible to conclude that ROS production may be additive, e.g. combining non-physiological oxygen tension with ambient light exposure might together jeopardize the development of human embryos *in vitro* (Noda *et al.*, 1994).

Oocytes and embryos have some capacity for intrinsic DNA repair, but the sperm cell does not (Ménézo *et al.*, 2007; Ménézo *et al.*, 2010). Protecting cells *in vitro* – sperm, oocytes, and embryos – involves minimizing environmental stressors; the culture environment contributes to oxidative damage, and inclusion of antioxidants in the medium may provide additional protection (Bavister, 2000; Guerin *et al.*, 2001; Martín-Romero *et al.*, 2008; Moshkdanian *et al.*, 2011; Orsi & Leese, 2001). Altering the ability of the oocyte and embryo to repair DNA that has been damaged by exposure to increased ROS can lead to altered transcription of repair mechanisms, and potentially reduced fertility (El-Mouatassim *et al.*, 2007).

Therefore, culture system design should focus on minimizing opportunities for additional ROS production. A systematic approach to designing an embryo culture system includes choosing from the many commercial culture media available (although some clinics still manufacture their own culture media in-house); this is one of the more critical aspects of human clinical IVF. Culture media for human embryos have evolved significantly over the years, and yet unsubstantiated and often misinterpreted concepts regarding culture medium components and phototoxicity have been perpetuated over the years, stemming from several early tumor cell culture studies. For example, Spierenburg's team found that cell culture media that were not formulated for use with mammalian embryos *in vitro* containing HEPES and riboflavin compounds (B_2 vitamins), but not pyruvate, were vulnerable to fluorescent light-induced peroxidation in the presence of oxygen (Spierenburg *et al.*, 1984). The protective role of pyruvate against downstream light-induced peroxidation toxicity in culture media may be via its action as an oxidative scavenger, thus maintaining intracellular glutathione concentrations (Babich *et al.*, 2009).

Current culture media for human and other mammalian embryos typically contain pyruvate; some contain some or all essential and non-essential amino acids, but they do not usually contain riboflavin compounds, which are responsible for light-induced phototoxicity (Edwards & Silva, 2001). However, significant ROS generation may still occur simply by warming the medium to incubator temperatures (Martín-Romero *et al.*, 2008).

Quality of the oil used as a culture medium overlay is an additional important and often overlooked component of the culture system. A commercial oil product may arrive sterile from the manufacturer with a valid certificate of analysis, but improper transportation temperature, exposure to sunlight and/or visible light or improper storage may result in adulteration of the oil (see Chapter 3). Peroxides can form in a cascade reaction as a result of exposure to light and/or heat, with transfer of water soluble contaminants to the culture drops (Otsuki *et al.*, 2009, 2007; Provo & Herr, 1998).

It is clear that ROS generation *in vitro* is unavoidable; addition of exogenous antioxidants (e.g. ascorbic acid), and supporting gamete and embryonic health to drive intrinsic antioxidant capabilities may counter the more damaging effects of ROS exposure. Technologies already exist to limit external stressors, for example, supraphysiological oxygen concentrations can be avoided by using a specific incubator design to reduce oxygen concentrations, and pre-mixed low oxygen concentration gases are available for use with

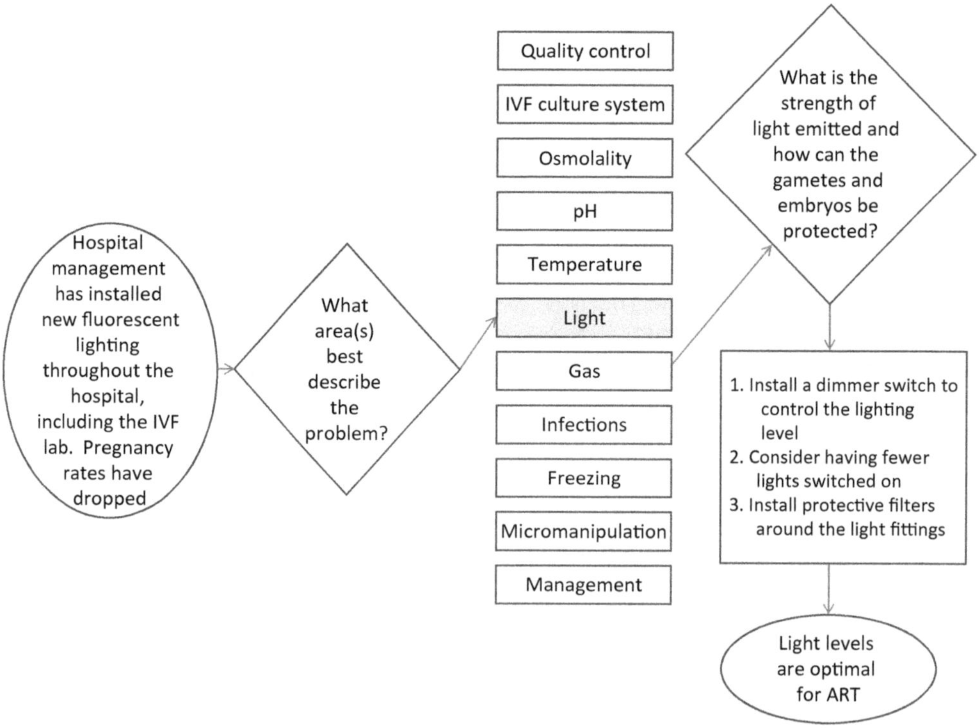

Fig. 7.2 Process map for troubleshooting light exposure problems.

mini-incubators and enclosed containers. Blue wavelength light energies can be avoided altogether by selecting lighting sources and/or use of light filtration devices; exposure of gametes and embryos to any light energies can be minimized by limiting the time that these cells are maintained outside of the incubator. Careful selection and storage of embryo culture media and oil for overlays can contribute to reducing ROS exposure. Nonetheless, even if damage from ROS is avoided via some of the methods above, damage from light that is not mediated by ROS is still possible.

Conclusion

No conclusive data confirms that light is harmful to human gametes or embryos, but there is substantial evidence that light is harmful to non-primate mammalian gametes and

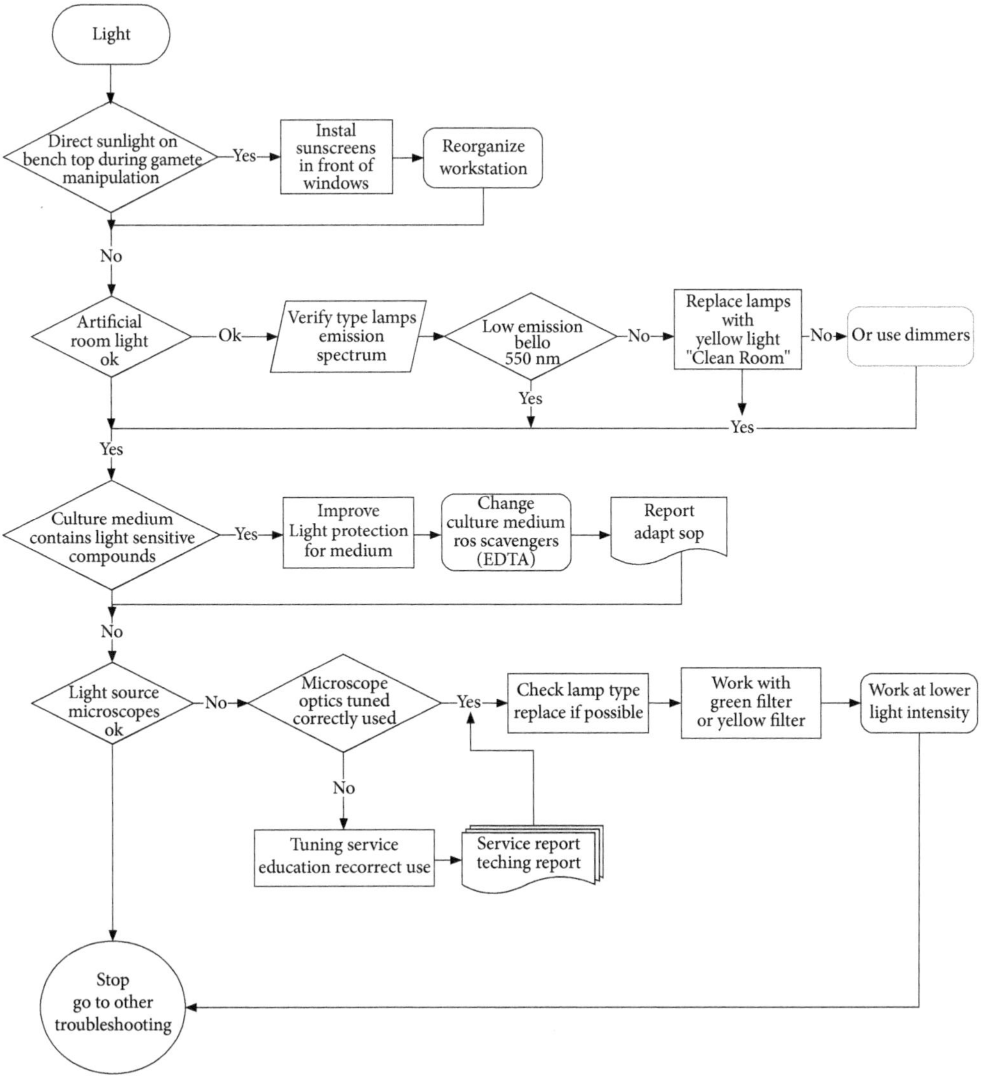

Fig. 7.3 Flowchart for troubleshooting light exposure problems.

embryos. It is also known that light can affect the quality of oil and culture media, including buffers such as HEPES. Light in the blue visible and ultraviolet spectrum (<500 nm) appears to have the highest potential for harm. There may be ways to reduce the effects of light on embryos and media by the inclusion of antioxidants or the exclusion of photo-oxidative media components. Light exposure can also be reduced by reducing the amount of harmful wavelengths in our laboratories via limiting exposure to any light, using ambient light filters, and avoiding most fluorescent lighting.

Embryology laboratories should not be located in areas where direct sunlight might cause damage. Care should be taken with hood lights, ambient lights, headlamps, and microscope lamps. A green bypass filter may be prudent when viewing gametes and embryos. With new methods for determining embryo quality and suitability for transfer with frequent monitoring using in-incubator optics, it is important to try to increase our understanding of the role light might play in the growth and development of human embryos.

References

Agarwal A, Allamaneni SSR (2004). Oxidants and antioxidants in human fertility. *MEFS J*, **9**, 187–97.

Agarwal A, Said TM, Bedaiwy MA, Banerjee J, Alvarez JG (2006). Oxidative stress in an assisted reproductive techniques setting. *Fertil Steril*, **86**, 503–12.

Babich H, Liebling EJ, Burger RF, Zuckerbraun HL, Schuck AG (2009). Choice of DMEM, formulated with or without pyruvate, plays an important role in assessing the *in vitro* cytotoxicity of oxidants and prooxidant nutraceuticals. *In Vitro Cell Dev Biol Anim* **45**, 226–33.

Bavister BD (2000). Interactions between embryos and the culture milieu. *Theriogenology*, **53**, 619–26.

Bedford JM, Dobrenis A (1989). Light exposure of oocytes and pregnancy rates after their transfer in the rabbit. *J Reprod Fertili*, **85**, 477–81.

Catt JW, Henman M (2000). Toxic effects of oxygen on human embryo development. *Hum Reprod*, **15** Suppl 2, 199–206.

Daniel JC (1964). Cleavage of mammalian ova inhibited by visible light. *Nature*, **201**, 316–17.

Del Giudice M (2011). Alone in the dark? Modeling the conditions for visual experience in human fetuses. *Dev Psychobiol*, **53**, 214–19.

Edwards AM, Silva E, (2001). Effect of visible light on selected enzymes, vitamins and amino acids. *J Photochem Photobiol B, Biol*, **63**, 126–31.

Edwards RG (1998). New concepts in embryonic growth and implantation. *Hum Reprod*, **13**(Suppl 3), 271–82.

Evans J, Wells C, Hood K (1999). A possible effect of different light sources on pregnancy rates following gamete intra-Fallopian transfer. *Hum Reprod*, **14**(1), 80–2.

El-Mouatassim S, Bilotto S, Russo GL, Tosti E, Menezo Y (2007). APEX/Ref-1 (apurinic/apyrimidic endonuclease DNA-repair gene) expression in human and ascidian (*Ciona intestinalis*) gametes and embryos. *Mol Hum Reprod*, **13**, 549–56.

Grzelak A, Rychlik B, Bartosz G (2001). Light-dependent generation of reactive oxygen species in cell culture media. *Free Radic Biol Med*, **30**, 1418–25.

Guerin P, Mouatassim SE, Menezo Y (2001). Oxidative stress and protection against reactive oxygen species in the pre-implantation embryo and its surroundings. *Hum Reprod Update*, **7**, 175–89.

Halliwell B (2006). Reactive species and antioxidants. Redox biology is a fundamental theme of aerobic life. *Plant Physiol*, **141**, 312–22.

Hill Jr. RB, Bensch KG, King DW (1960a). Photosensitization of nucleic acids and proteins: the photodynamic action of acridine orange on living cells in culture. *Exp Cell Res*, **21**, 106–17.

Hill RB Jr., Bensch KG, King DW (1960b). Photosensitization of nucleic acids and proteins: the photodynamic action of acridine orange on living cells in culture. *Exp Cell Res*, **21**, 106–17.

Hirao Y, Yanagimachi R (1978). Detrimental effect of visible light on meiosis of mammalian eggs *in vitro*. *J Exp Zool*, **206**, 365–9.

Jacques SL, Weaver DR, Reppert SM (1987). Penetration of light into the uterus of pregnant mammals. *Photochem Photobiol*, **45**, 637–41.

Korhonen K, Sjövall S, Viitanen J, Ketoja E, Makarevich A, Peippo J (2009). Viability of bovine embryos following exposure to the green filtered or wider bandwidth light during in vitro embryo production. *Hum Reprod*, **24**, 308–14.

Lavi R, Ankri R, Sinyakov M, *et al.* (2012). The plasma membrane is involved in the visible light–tissue interaction. *Photomed Laser Surg*, **30**, 14–19.

Lesser MP, Kruse VA, Barry TM (2003). Exposure to ultraviolet radiation causes apoptosis in developing sea urchin embryos. *J Exp Biol*, **206**, 4097–103.

Martín-Romero FJ, Miguel-Lasobras EM, Domínguez-Arroyo JA, González-Carrera E, Álvarez IS (2008). Contribution of culture media to oxidative stress and its effect on human oocytes. *RBM Online*, **17**, 652–61.

Ménézo Y, Dale B, Cohen M (2010). DNA damage and repair in human oocytes and embryos: a review. *Zygote*, **18**, 357–65.

Menezo Y Jr., Russo G, Tosti E, El Mouatassim S, Benkhalifa M (2007). Expression profile of genes coding for DNA repair in human oocytes using pangenomic microarrays, with a special focus on ROS linked decays. *J Assist Reprod Genet*, **24**, 513–20.

Moshkdanian G, Nematollahi-mahani SN, Pouya F, Nematollahi-mahani A (2011). Antioxidants rescue stressed embryos at a rate comparable with co-culturing of embryos with human umbilical cord mesenchymal cells. *J Assist Reprod Genet*, **28**, 343–9.

Noda Y, Goto Y, Umaoka Y, Shiotani M, Nakayama T, Mori T (1994). Culture of human embryos in alpha modification of Eagle's medium under low oxygen tension and low illumination. *Fertil Steril*, **62**, 1022–7.

Oh SJ, Gong SP, Lee ST, Lee EJ, Lim JM (2007). Light intensity and wavelength during embryo manipulation are important factors for maintaining viability of preimplantation embryos *in vitro*. *Fertil Steril*, **88**, 1150–7.

Orsi NM, Leese HJ (2001). Protection against reactive oxygen species during mouse preimplantation embryo development: role of EDTA, oxygen tension, catalase, superoxide dismutase and pyruvate. *Mol Reprod Dev*, **59**, 44–53.

Otsuki J, Nagai Y, Chiba K (2007). Peroxidation of mineral oil used in droplet culture is detrimental to fertilization and embryo development. *Fertil Steril*, **88**, 741–3.

Otsuki J, Nagai Y, Chiba K (2009). Damage of embryo development caused by peroxidized mineral oil and its association with albumin in culture. *Fertil Steril*, **91**, 1745–9.

Ottosen LDM, Hindkjær J, Ingerslev J (2007). Light exposure of the ovum and preimplantation embryo during ART procedures. *J Assist Reprod Genet*, **24**, 99–103.

Parr AC (2001). The candela and photometric and radiometric measurements. National Institute of Standards and Technology.

Provo MB, Herr C (1998). Washed paraffin oil becomes toxic to mouse embryos upon exposure to sunlight. *Theriogenology*, **49**, 214.

Schumacher A, Fischer B (1988). Influence of visible light and room temperature on cell proliferation in preimplantation rabbit embryos. *J Reprod Fertil*, **84**, 197–204.

Spierenburg GT, Oerlemans FT, Van Laarhoven JP, De Bruyn CH (1984). Phototoxicity of N-2-hydroxyethylpiperazine-N'-2-ethanesulfonic acid-buffered culture media for human leukemic cell lines. *Cancer Res*, **44**, 2253–4.

Stoien JD, Wang RJ (1974). Effect of near-ultraviolet and visible light on mammalian cells in culture II. Formation of toxic photoproducts in tissue culture medium by blacklight. *Proc Natl Acad Sci USA*, **71**, 3961–5.

Swain JE (2014). Decisions for the IVF laboratory: comparative analysis of embryo culture incubators. *RBM Online*, **28**(5), 535–47.

Takenaka M, Horiuchi T, Yanagimachi R (2007). Effects of light on development of mammalian zygotes. *Proc Natl Acad Sci USA*, **104**, 14289–93.

Umaoka Y, Noda Y, Nakayama T, Narimoto K, Mori T, Iritani A (1992). Effect of visual light on *in vitro* embryonic development in the hamster. *Theriogenology*, **38**, 1043–4.

Wang RJ (1975). Lethal effect of "daylight" fluorescent light on human cells in tissue-culture medium. *Photochem Photobiol*, **21**, 373–5.

Wang RJ, Nixon BR (1978). Identification of hydrogen peroxide as a photoproduct toxic to human cells in tissue-culture medium irradiated with "daylight" fluorescent light. *In Vitro*, **14**, 715–22.

Yamauchi Y, Yanagimachi R, Horiuchi T (2002). Full-term development of golden hamster oocytes following intracytoplasmic sperm head injection. *Biol Reprod*, **67**, 534–9.

Yang HW, Hwang KJ, Kwon HC, Kim HS, Choi KW, Oh KS (1998). Detection of reactive oxygen species (ROS) and apoptosis in human fragmented embryos. *Hum Reprod*, **13**, 998–1002.

Zan-Bar T, Bartoov B, Segal R, *et al.* (2005). Influence of visible light and ultraviolet irradiation on motility and fertility of mammalian and fish sperm. *Photomed Laser Surg*, **23**, 549–55.

Zieger MA, Glofcheski DJ, Lepock JR, Kruuv J (1991). Factors influencing survival of mammalian cells exposed to hypothermia. V. Effects of HEPES, free radicals, and H_2O_2 under light and dark conditions. *Cryobiology*, **28**, 8–17.

Zigler J, Lepe-Zuniga J, Vistica B, Gery I (1985). Analysis of the cytotoxic effects of light-exposed HEPES-containing culture medium. *In Vitro Cell Dev Biol - Plant*, **21**, 282–7.

Chapter 8

Gases, air quality, and volatile organic carbons (VOCs)

Gases

Gas supplies

A safe incubator environment for gametes and embryos depends crucially on a reliable supply of gases and gas mixtures of guaranteed consistency and quality. The principal gas mixtures used are "atmospheric," 5% carbon dioxide (CO_2) in air (20% oxygen, O_2) and "physiological/Trigas," 5%–6% CO_2/5% O_2/90% nitrogen (N_2). Commercially supplied gases have standard labels that indicate their purity and specifications, and these must comply with international standards, e.g. US Pharmacopeia (USP), European Pharmacopeia (EP), or Japanese Pharmacopeia (JP). Gases such as N_2 and O_2 are usually prepared by cryogenic liquefaction of ambient air, which is then distilled into separate fractions of various levels of purity. CO_2 may be obtained through processes in the petrochemical industry, or as a by-product of burning natural gas in a co-generation (combined heat and power) process. 100% purity cannot be achieved with any of these methods.

"Purity" of a gas is the term used by manufacturers to indicate the amount of other gases present, i.e. the absence of impurity. An "impurity" is any material or substance in the finished product that differs from the chemical composition of the requested compound. High purity means a low amount of other gases, and high purity gases are considered to be of better quality (and are usually more expensive). Gas purity is expressed either as a percentage value, or as a decimal fraction that represents an abbreviation of the percentage value: the first digit indicates the number of 9s in the percentage value, and the last digit represents the last digit of the percentage value (e.g. 2.5 = 99.5%). Impurity levels can also be reported as parts per million (ppm) (Table 8.1).

Gas cylinders are supplied with a Certificate of Analysis that complies with ISO6141, the standard that defines the information required for certificates of analysis (ISO6141:2000). The certificate confirms that the maximum limits for impurities have not been exceeded, and contains information that allows the supplied gas to be fully traced back to the raw material. Certification can be provided either by weight (gravimetric certification, careful measurement of the weight of the gases being filled into the cylinder) or by analyzing the gas from a cylinder and comparing the result to a known gas standard: certificates normally indicate which standard is used (Table 8.2). However, it is important to be aware that standards for supplies of compressed gases are based upon criteria that are not designed for cultured and unprotected cells in an IVF system.

Table 8.1 Gas purity levels reported as % or as parts per million (ppm)

Purity	Level	Reported impurity level as%	Reported impurity as ppm
99.0%	2.0	1%	10 000 ppm
99.5%	2.5	0.5%	5 000 ppm
99.9%	3.0	0.1%	1 000 ppm
99.99%	4.0	0.01%	100 ppm
99.999%	5.0	0.001%	10 ppm
99.9995%	5.5	0.0005%	5 ppm
99.9999%	6.0	0.0001%	1 ppm (1000ppb)

Table 8.2 Standards used for gas purity certification

Alpha standard	Certified value within +/− 1% of the actual concentration value, e.g. for an analytical result of 10% CO_2 in N_2 the CO_2 will be between 9.9% and 10.1%.
Beta standard	Certified value within +/− 2% of the actual concentration value e.g. for an analytical result of 10% CO_2 in N_2 the CO_2 will be between 9.8% and 10.2%.
Certified mixture	Individual cylinder analysis certified to be within +/− 2% to +/− 5% of the actual concentration value.
Uncertified mixture	Individual or batch analysis may be performed but no analyzed value is quoted.

Regardless of certified "purity", gas bottles can still be a source of other pollutants such as volatile organic carbons (VOCs) or trace amounts of heavy metals. Table 8.3 shows a typical analysis for Grade 3.0 liquid CO_2.

Monitoring gas mixtures

Incubator CO_2 and O_2 levels can be measured/monitored by physical or chemical methods. Any external electronic device used for CO_2/O_2 monitoring requires regular calibration, with an appropriate certificate delivered for each calibration. The calibration certificate must include information about the uncertainty range of measurements.

CO_2 measurement

Physical methods

Physical methods are currently more popular, and can be highly reliable provided that accurate calibration is carried out.

(a) Electrical resistance (thermal conductivity, TC) is used in the older generation of incubators; the resistance between two sensors (one inside the incubator and one enclosed) is related to the amount of CO_2, but depends upon the humidity level in the chamber: >90% humidity is required for precise function, and this is also affected by temperature changes. The monitor must be calibrated in a water-saturated atmosphere,

(a) (b)

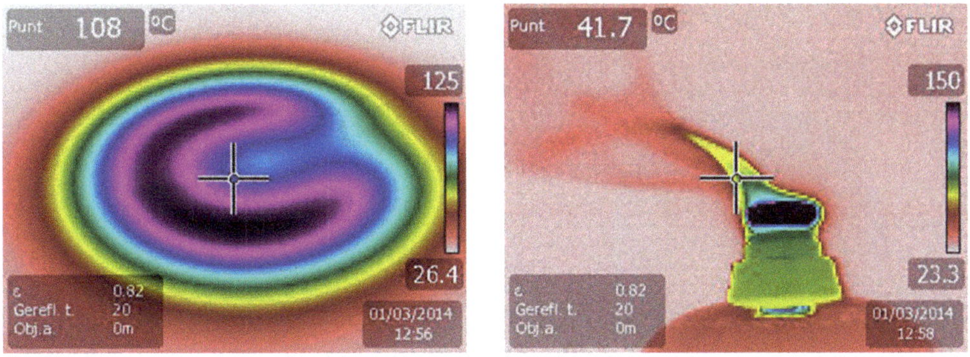

Fig. 6.1 Photo-thermographs that illustrate heat conduction and convection. (a) Heat conduction from an electric cooker hob plate; (b) Heat convection from an oil burner. *(Quartec GmbH.)*

(a) (b)

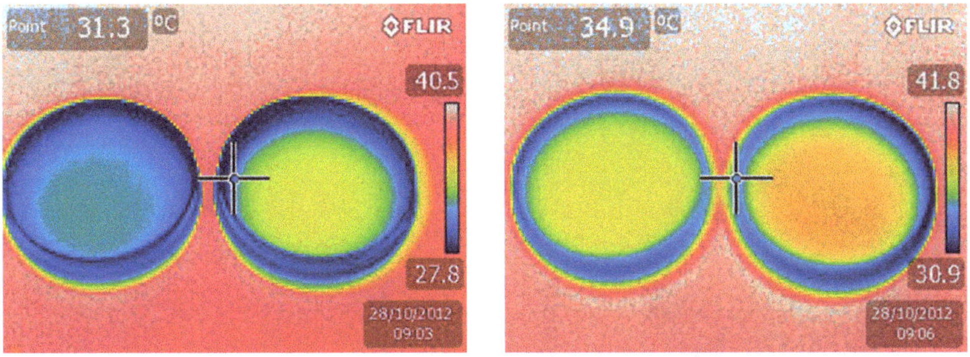

Fig. 6.4 Two Petri dishes filled with oil at the same temperature were placed on a heating stage set to 37 °C. A preheated Thermocoin (LivingCell) was placed under the dish on the right, and thermographs taken at 3-minute intervals. After 3 minutes (9.03) the dish on the left (a) had lost heat compared with the dish on the right (b). It took a further 3 minutes (9.06) for the dish without Thermocoin to reach the same temperature as the dish with the coin. The blue circles at the perimeter of both dishes indicate a lower temperature – a factor to be considered when positioning drops in an ICSI dish. *(Quartec GmbH.)*

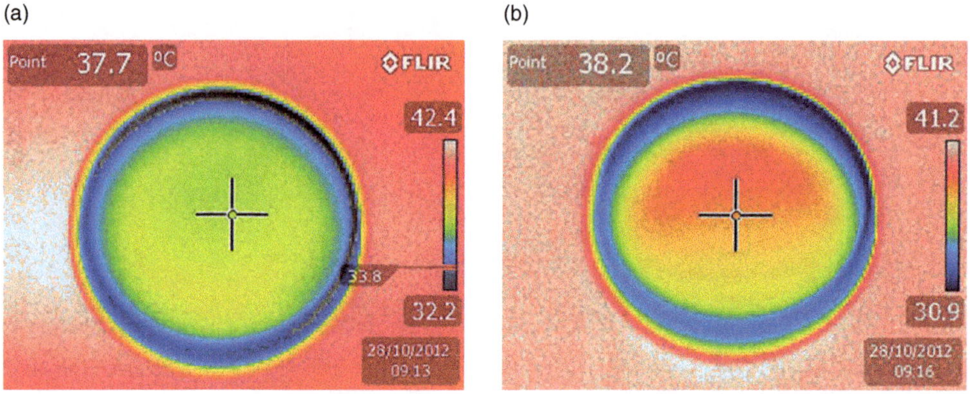

Fig. 6.5 Thermographs taken of the same dish, to illustrate artefacts that may influence interpretation of results. (a) Photo taken at a correct angle. (b) Photo taken with the camera positioned incorrectly; the angle of incidence creates an artifact, so that the upper part of the dish appears to be hotter, when it is not. A similar artifact can be created when light is reflected from a surface that is being measured. *(Quartec GmbH.)*

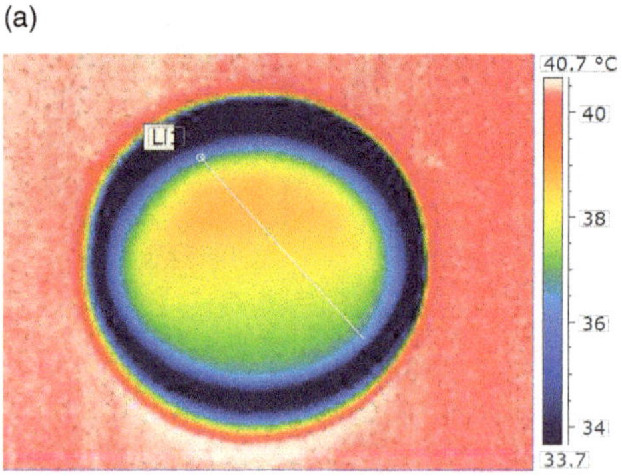

Fig. 6.9 Demonstration of unequal distribution of temperature in an ICSI dish. (a) Thermograph shows that the blue rim is colder than the center of the dish.

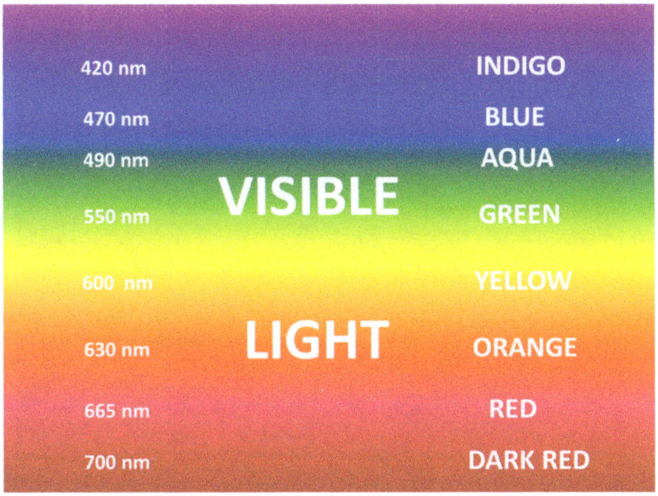

Fig. 7.1 The spectrum of visible light; numbers on the left represent wavelength in nanometers. Ultraviolet light has wavelengths between approximately 310 nm and 100 nm, with photon energies from 4 eV to 12.4 eV. Wavelengths in the 400–300 nm range are known as "near ultraviolet" (visible to birds, insects and fish), and have photon energies of 3.10–4.13 eV. At the other end of the spectrum, infrared (IR) light wavelengths extend from the red edge of the visible spectrum at 700 nm to 1 mm, with photon energies 1.24 mEV–1.7 eV.
Most of the thermal radiation emitted by objects near room temperature is infrared radiant energy.

Fig. 8.2 Probes used to monitor incubators. (1) A capacitive humidity probe consists of a hygroscopic dielectric material placed between a pair of electrodes to form a small capacitor. The probe contains a dielectric material (usually plastic or polymer) with a dielectric constant ranging from 2 to 15: when no moisture is present in the sensor, both this constant and the sensor geometry determine the value of capacitance. (2) Temperature probes consist of two wire class A probes. (3) The CO_2 sampling system transports a gas sample from the incubator atmosphere to a centralized measuring station (often fitted into the lab ceiling). Samples from several incubators can be analyzed by a single CO_2 sensor within a CO_2 substation. Sample channels can be measured every 20 minutes. Excess humidity is automatically removed by moisture separators. The internal CO_2 sensor is automatically calibrated using outside air (0% CO_2) and a calibration gas supply. *(Image kindly provided by the Hewitt Fertility Centre, Liverpool Women's Hospital, UK. Information kindly provided by Mr. Brian Lewin, Sedna Scientific, Dronfield, Derbyshire.)*

(a)

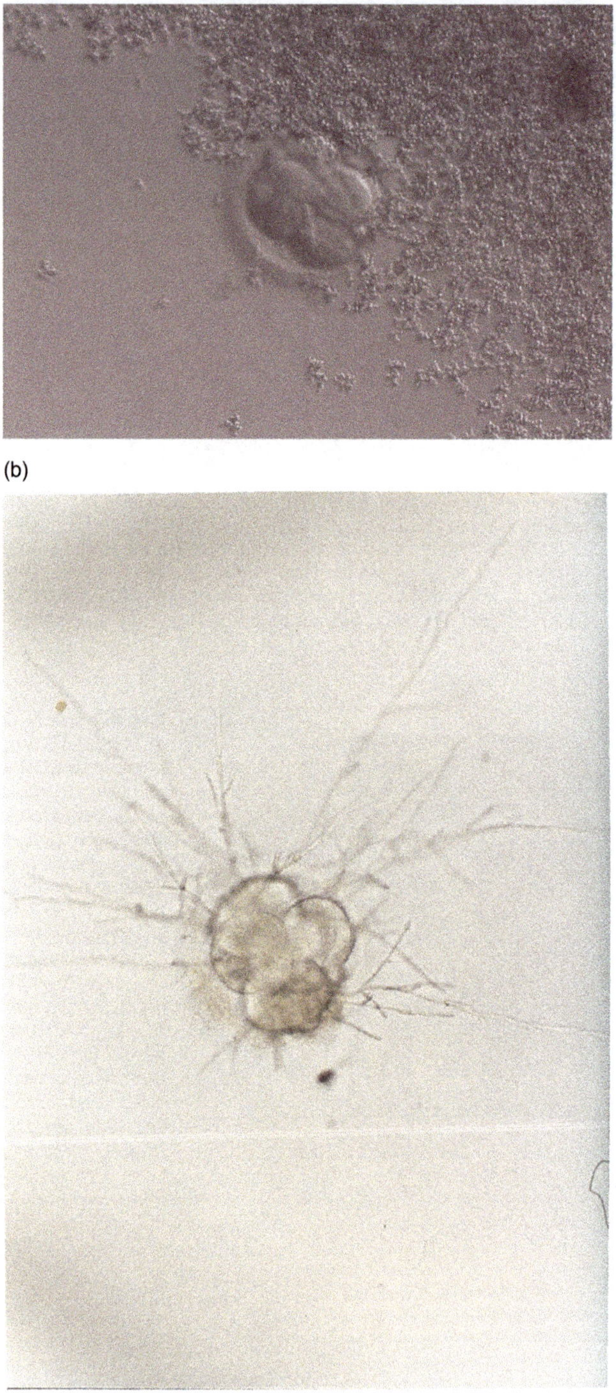

(b)

Fig. 9.2 Human embryos with yeast (*C. albicans*) contamination in the culture dish. (a) Embryo with spherical budding cells (blastoconidia) in the medium. (b) Embryo surrounded by filamentous budding elongated cells (pseudohyphae).

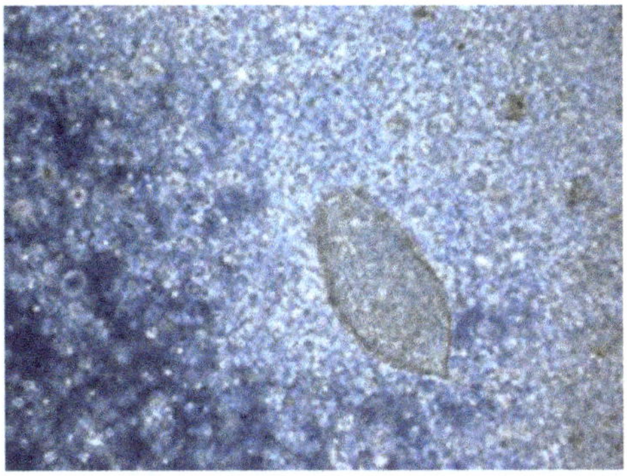

Fig. 9.4 *Schistosoma haematobium* observed in a wet slide preparation during a diagnostic semen analysis (image taken at 400x). (*Photomicrograph by Sue Kenworthy, Lead Andrologist, taken at Portsmouth Hospitals NHS Trust, UK.*)

(a)

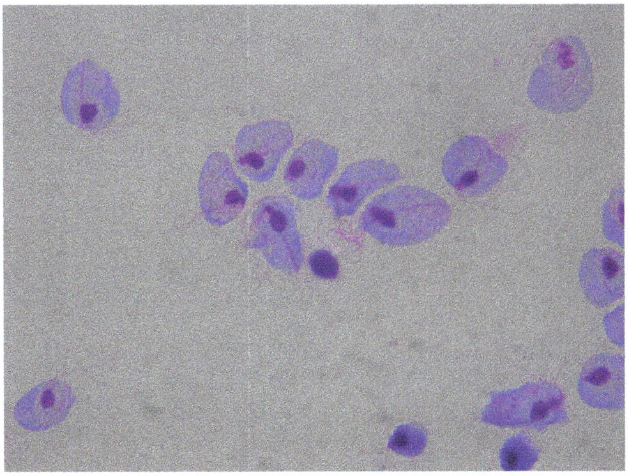

(b)

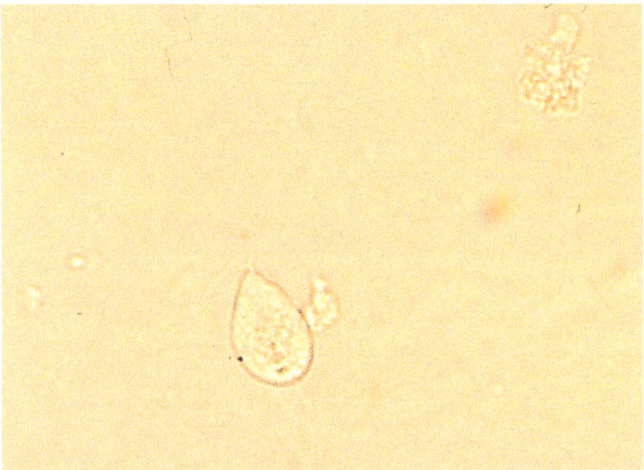

Fig. 9.5 *Trichomonas vaginalis.* (a) Specimen cultured in Diamond's Medium stained with Wright's Giemsa 100x. *T. vaginalis* flagellates are pyriform, round or oval in shape, and measure 7–30 by 5–15 μM. Each organism has a single anterior nucleus and a tuft of four anterior flagella and one short posterior flagellum. A linear axostyle extends posteriorly and projects out of the organism. (b) *T. vaginalis* in a wet mount from urine, unstained 100x, illustrating the pyriform shape and anterior flagellae. (*Images kindly provided by Julie A Ribes, MD, PhD, Director of Clinical Microbiology, University of Kentucky Hospital, Lexington, KY.*)

(a)

(b)

Fig. 10.3 Cryostore management. (a) Colored goblets help in keeping track of samples. They can be arranged in alphabetical order, as shown here: aqua, black, blue, brown, green, gray, orange, pink, purple, tangerine, yellow, or in the order of the rainbow/white light spectrum. (b) A canister that has just been removed from liquid nitrogen, containing straws that are distributed between different colored goblets. (*Images kindly provided by the Hewitt Fertility Centre, Liverpool Women's Hospital, UK.*)

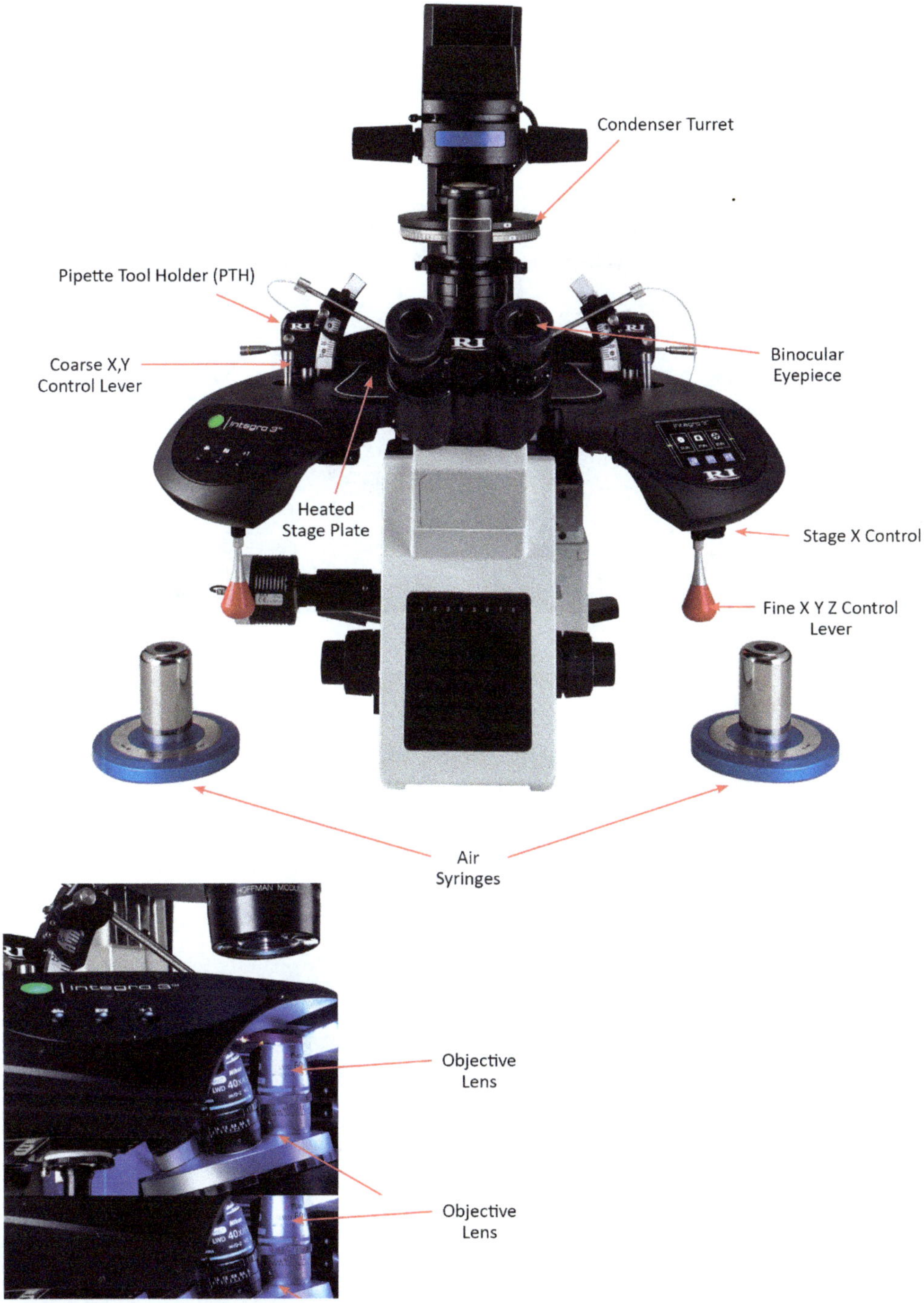

Fig. 11.1 Micromanipulation station (*Integra 3*^{TM}, Research Instruments Ltd), with parts identified. (*Image kindly supplied by Research Instruments Ltd, UK. www.research-instruments.com.)*

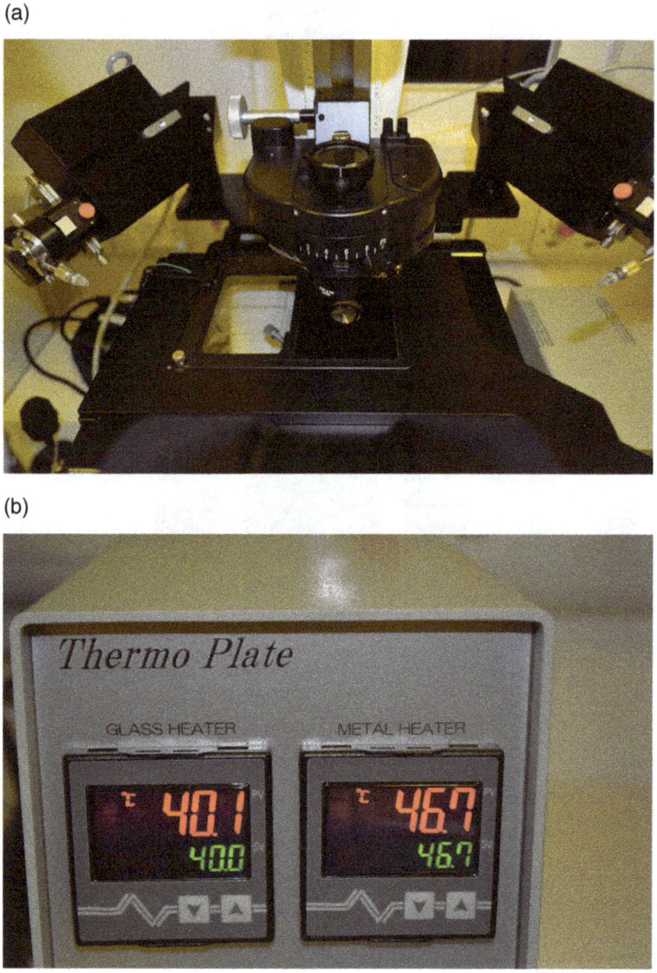

Fig. 11.2 Different types of heated stage may be required for some micromanipulation activities. (a) Micromanipulation system with two heated stages, glass on the left, and metal on the right, with an aperture for the microscope objective. (b) Each stage has a separate temperature controller, which should be validated to ensure that the microdroplet temperature in a dish placed on each stage is maintained at 37 °C. This usually requires different settings for the dual heated stage controller, depending on the ambient air temperature. (*Images kindly provided by the Hewitt Fertility Centre, Liverpool Women's Hospital, UK.*)

Table 8.3 Grade 3.0 liquid carbon dioxide specification

Component	Concentration
Carbon dioxide	99.9% v/v min
Moisture	50 ppm v/v max (20 ppm w/w max.)
Acidity	To pass JECFA or ISBT test
Ammonia	2.5 ppm v/v max
Oxygen	30 ppm v/v max.
Oxides of nitrogen (NO/NO_2)	2.5 ppm v/v max. each
Non-volatile residue (particulates)	10 ppm w/w max.
Non-volatile organic residue (oil and grease)	5 ppm w/w max.
Phosphine***	≤= 0.3 ppm v/v
Total volatile hydrocarbons (calculated as methane)	50 ppm v/v max. of which 20 ppm v/v max non-methane hydrocarbons.
Acetaldehyde	0.2 ppm v/v max.
Benzene	0.02 ppm v/v max.
Carbon monoxide	10 ppm v/v max.
Methanol	10 ppm v/v max.
Hydrogen cyanide*	<0.5 ppm v/v
Total sulfur (as S)**	0.1 ppm v/v max.
Taste and odor in water	No foreign taste or odor

* Analysis necessary only for carbon dioxide from coal gasification sources.
** If the total sulfur content exceeds 0.1 ppm v/v as sulfur then the species must be determined separately and the following limits apply:
Carbonyl sulfide 0.1 ppm v/v max.
Sulphur dioxide 1.0 ppm v/v max.
Hydrogen sulfide 0.1 ppm v/v max.
*** Analysis necessary only for carbon dioxide from phosphate rock sources.
This conforms to BS4105 Type 2 for food use.
(Courtesy of Mr Nick Brown, Air Products, Worcester, UK.)

and if the conditions of temperature or humidity inside the chamber change (e.g. after door opening), the sensor cannot give an accurate reading until normal conditions are re-established. This can take 10–15 minutes after door opening/closing, according to the size of the incubator, the amount of water in the humidity pan, etc. If the water pan dries out, CO_2 can no longer be measured accurately.

(b) Infrared (IR) sensors use an optical sensor: IR light is emitted, which is absorbed by CO_2 in a linear fashion, and the sensor determines the amount of CO_2 present according to the amount absorbed by the IR light. This type of sensor is more expensive and has a limited lifespan, but readings don't require humidity, and CO_2 readings recover more quickly after door opening.

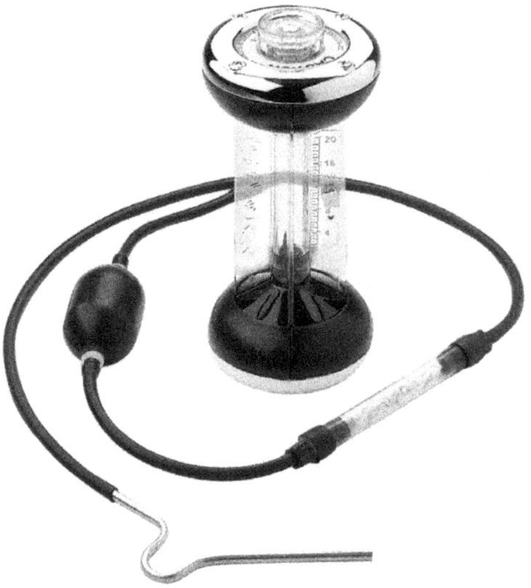

Fig. 8.1 Bacharach gas analyzer; the pink fluid contains potassium hydroxide with a red dye added for visibility. (*Image kindly provided by Bacharach Inc, USA.*)

Chemical methods

(a) *"Orsat Gas Analyzer,"* better known as the *"Fyrite-Bacharach" analyzer* (Fig. 8.1). A manual pump is used to draw gas for analysis into a burette containing a small amount of colored fluid, and the burette is inverted several times to mix the gas and fluid. For ease of calculation, a volume of 100 ml of gas is typically withdrawn. The stopcocks that isolate the absorption burette are used to level the gas in the leveling bottle, and the burette is adjusted to its zero point either by opening a valve or by lowering the scale ruler. This operation is repeated before each measurement. The volume of gas remaining in the burette indicates the percentage of CO_2 absorbed. The same technique can be used with pyrogallol to measure oxygen, and with ammoniacal copper chloride to measure carbon monoxide. This method has several pitfalls:

  (i) The volume of fluid supplied with the device is sufficient for only approximately 350 measurements, and there is no clear indication when it has been exhausted.
  (ii) The rubber bulb of the pump must be squeezed 18 times in order to sample 100 ml of gas: an inevitable source of error.
  (iii) The reaction fluids used in the burette are corrosive, toxic, and a source of VOCs: the fluid used in a Fyrite analyzer is potassium hydroxide, a toxic alkaline solution with a red dye added for visibility. These instruments must be handled with great care.

(b) "*Draeger Tubes*". This method also uses a manual pump to withdraw gas, and is based on a reaction between CO_2 and a solid chemical, with change in color from white to blue (methyl violet):

$$CO_2 + N_2H_2 - - - \longrightarrow NH_2\text{-}NH\text{-}COOH$$

The Draeger AccuroPump hand pump has a fixed volume of 100 ml – repeated pumping is not necessary. However, leaks in the tubing and a large dead space in the tubing that connects the pump to the incubator are a source of imprecision: the smallest errors are still within +/− 0.5%.

Fig. 8.2 Probes used to monitor incubators. (1) A capacitive humidity probe consists of a hygroscopic dielectric material placed between a pair of electrodes to form a small capacitor. The probe contains a dielectric material (usually plastic or polymer) with a dielectric constant ranging from 2 to 15: when no moisture is present in the sensor, both this constant and the sensor geometry determine the value of capacitance. (2) Temperature probes consist of two wire class A probes. (3) The CO_2 sampling system transports a gas sample from the incubator atmosphere to a centralized measuring station (often fitted into the lab ceiling). Samples from several incubators can be analyzed by a single CO_2 sensor within a CO_2 substation. Sample channels can be measured every 20 minutes. Excess humidity is automatically removed by moisture separators. The internal CO_2 sensor is automatically calibrated using outside air (0% CO_2) and a calibration gas supply. A black and white version of this figure will appear in some formats. For the color version, please refer to the plate section. *(Image kindly provided by the Hewitt Fertility Centre, Liverpool Women's Hospital, UK. Information kindly provided by Mr. Brian Lewin, Sedna Scientific, Dronfield, Derbyshire, UK.)*

Oxygen measurement

Three types of oxygen sensors are used in incubators: zirconia (zirconium oxide), galvanic (paramagnetic), and Clark electrode (Fig. 8.2).

1. *Zirconia* sensors are solid-state, are very precise and don't require continuous calibration. They measure the difference in oxygen content between the incubator gas and external air, generating a voltage/resistance change according to the difference between the two. The signal output is proportional to the partial pressure of oxygen in the sample gas, with a linear relationship. The sensor element is a ceramic cylinder plated with porous platinum electrodes on the inside and outside. The whole assembly is protected by a metal gauze. Zirconium O_2 sensors are usually found on the back wall of an incubator. The porous surface can become contaminated with micro-organisms, and is difficult to decontaminate. If the incubator is powered off with humidity still inside the chamber, they can develop condensation that will affect the function of the sensor.
2. *The Clark electrode* measures oxygen on a catalytic platinum surface, using the net reaction

   $$O_2 + 4e^- + 2H_2O - -\!\!\rightarrow 4OH^-$$

   The measuring principle is based on an electrical current passing through a cathode and an anode submerged in an electrolyte solution. O_2 enters the sensor through a permeable Teflon membrane by diffusion, and is reduced at the cathode, creating an electrical current that can be measured. There is a linear relationship between the O_2 concentration and the electrical current; O_2 in the sample can be measured after two-point calibration with 0% and 100% air saturation. This type of sensor was very common in the first generation of incubators; regular replacement of the membrane required skilful operation, and the glass electrolyte containers had a tendency to leak frequently. Evaporation of the electrolyte created an additional problem, with formation of a salt crust. The glass electrode was often broken whilst trying to replace the Teflon membrane.

3. *Galvanic* sensors work on the same principle as the Clark electrode, measuring O_2 via its reduction at a cathode. This is accomplished by using two different metals (lead and gold, or lead and silver) in the presence of an electrolyte, so that a voltage relative to O_2 concentration is produced. These sensors have a battery, and therefore the sensor must be calibrated periodically to check battery life.

Gas cylinder storage

Any errors in delivery of gas supplies to the laboratory can lead to serious problems, and all deliveries must be carefully checked. A secure and safe storage location for gas cylinders, both during use and in storage is essential: this includes all of the gases used in the laboratory, as well as those used by the medical team during any surgical procedure. The storage room, designated as a "No Smoking" area, should be as far as possible from any potential sources of heat or ignition. Different countries have local and national regulations for gas storage areas, which must be adhered to. Specifications generally include:

- Adequate ventilation, to prevent gas build-up in case of leakage.
- Free of clutter.
- Protected from extremes of weather, including high humidity (causing rust and corrosion) and extremes of temperature that will affect gas stability.
- Vertical cylinder storage, with the connections uppermost and secured to prevent them from falling, preferably with specific clamps or chains (Fig. 8.3).
- It is helpful to store the cylinders in order of delivery, so that the "oldest" cylinders are used first. Inspect each gas cylinder for damage at the time of delivery. It is also prudent to assess the number of cylinders stored at the IVF clinic at any one time, as most companies charge a rental for each cylinder after delivery. Careful assessment of gas usage and frequency of cylinder deliveries can result in significant cost savings: Trigas (5% CO_2/5% O_2/90% N_2) can be a very expensive gas mix. Valves must be closed when not in use, with valve protectors fitted to keep them in place.

Safety precautions for gas cylinders

Specification sheets for all gas cylinders contain details of safety aspects relating to the cylinder and the gas, including:

- Identification of the gas and the delivery company
- Known hazards of the gas
- Full composition of the gas supplied

Fig. 8.3 Example of inadequate gas tank storage area; these 100% CO_2 and 5% CO_2 cylinders were stored within an area of the laboratory, without adequate ventilation and no oxygen sensor to monitor in case of leaks. The cylinders are not stabilized, but held in place with loose chains only. *(Quartec GmbH.)*

- First aid and fire-fighting measures
- Accidental release measures
- Handling and storage
- Recommended PPE (Personal Protective Equipment)
- Physical and chemical properties
- Gas stability and reactivity
- Toxicological and ecological information
- Disposal, transport and regulatory information.

Asphyxiation is the most common hazard associated with improper use or leakage of gases, and for this reason storage rooms must be well ventilated, and preferably have a low O_2 monitor installed. Soapy water can be used to test for leaks. A person suffering from asphyxiation may feel dizzy, or even unconscious. If someone working with gases suddenly becomes dizzy, they should immediately stop what they are doing, and be escorted to an outside area away from the gases to recover and receive medical help. It is always advisable to work in pairs when dealing with gases.

Make sure that is safe to carry or move a gas cylinder before doing so, and wear protective gloves and footwear. A trolley or other device can be used to move cylinders, provided that the cylinder is secured, full training has been given and sloped access is available to avoid stairs. Never attempt to carry a gas cylinder unless you have had proper manual handling training and have carefully assessed the load. Any excessive shock during transit may damage the cylinder: if damage is suspected, contact the delivery company immediately to arrange for a replacement.

Make sure that the cylinder valve is protected at all times, usually by a protective guard; never lift cylinders using the valve guard.

Regulators and equipment for incubator gases

Different cylinders require different regulators; make sure that each type of cylinder is fitted with the correct regulator of appropriate size and pressure rating. The regulator fits directly onto the outlet valve of the cylinder via a screw thread – do not attach any additional tubing between regulator and cylinder! Turn the gas cylinder off fully and make sure that the regulator is fully "screwed out" before fitting, so that no gas flows through the regulator when the cylinder is turned on. Check that the cylinder is supplied with a non-return valve, so that the regulator can prevent backflow that might contaminate the gas supply.

Inspect regulators routinely to make sure that they are not old, and are fit for purpose. They should fit onto the cylinder thread with minimal resistance, and must not be 'adapted'. Never force a regulator onto a cylinder – if force is required, there is a problem. Do not use a regulator if the thread is loose-fitting or appears to be damaged. Use of polytetrafluoroethylene (PTFE) tape as a temporary connection is not reliable, and is not recommended. Never use an oil or lubricant to try to ease a regulator fitting if the thread is too tight – some gases may react violently with lubricants. If there is the slightest problem with a regulator, seek advice from the company that supplied the regulator, the cylinder, or the incubator.

A manifold can be used to connect several cylinders to a single regulator in order to supply an incubator with a mixture of gases (Fig. 8.4a). Two cylinders can be fixed to a common outlet via an auto-changeover manifold, so that if one cylinder empties, the regulator will switch to the fuller cylinder (Fig. 8.4b). The embryologist must be completely familiar with the function of the manifold, and ensure that one full cylinder is always connected. Switch the regulator manually after each change of empty cylinder, and check the system every day, inverting the handle in the middle. If this is neglected, both bottles will empty in parallel, leaving no "reserve" gas. Also check that the pressure inlet for the cylinder and regulator are correct for the incubator they are connected to, and that the correct gas has been supplied: some incubators can operate from either 100% CO_2 or 5% CO_2: using the wrong gas supply will have disastrous consequences! It is helpful to have a control dish containing media with phenol red indicator, as this will turn yellow in highly acidic conditions (e.g. a 100% CO_2 environment) and pink in alkaline conditions (e.g. an environment where there is no CO_2).

Ambient air quality

Air conditioning systems

Air cleanliness is classified according to the number and size of particles within a sample of air, measured in particles per cubic foot or cubic meter of air. The US Federal Standard 209E defines air quality based upon the maximum allowable number of particles 0.5 microns and larger per cubic foot of air: Class 1, 100, 1000, 10 000 and 100 000. The lower the number, the cleaner the air. The ISO 14644-1 classifications are defined as ISO Class 1 to 9 (ISO14644-1:1999). Cleanest, ultra-pure air is Class 1 (Table 8.4).

Laboratory air quality can be maintained by using positive air pressure relative to adjacent areas; adjacent rooms with different clean area classification should have a pressure differential of 10–15 pascals, with higher pressure in the most critical area. The position of the air inlet and the dynamics of the heating, ventilation, and air conditioning (HVAC) system are crucial. Unless the source of the air is very clean, an HVAC system can

(a) (b)

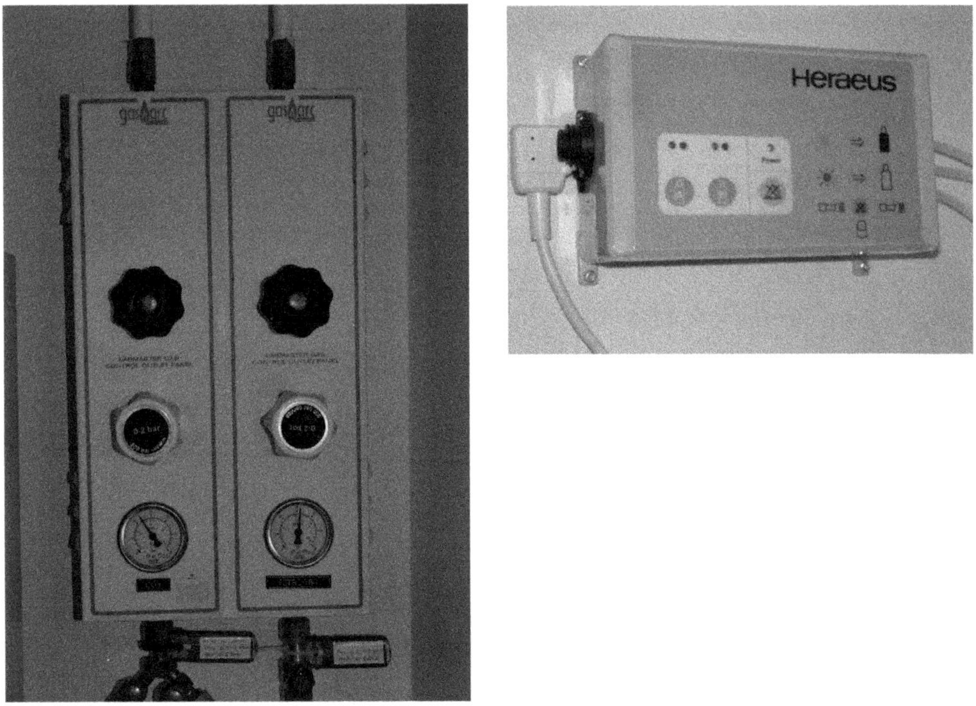

Fig. 8.4 Manifolds that allow two gas cylinders to be fixed to a common outlet. (a) Two regulators supplying purified gas (CO_2 gas on the left; N_2 gas on the right) to a low oxygen incubator. *Image kindly provided by the Hewitt Fertility Centre, Liverpool Women's Hospital, UK.* (b) Electronic auto-changeover manifold (detail of manifold fixed to the wall in Fig. 8.3). *(Quartec GmbH.)*

potentially increase the volume of dirty/toxic air brought into the laboratory. This may be particularly important with VOCs that may be embryotoxic.

The location of an IVF laboratory merits careful consideration, including the outside environment; any potential problems need to be assessed specifically for IVF rather than what may be considered "standard" in a hospital setting. Highly polluted air can be man-made (e.g. from traffic/industry/building works) or relate to geography (e.g. from air-borne particles such as desert sand). In such cases, the number of initial external air filters may need to be increased in order to reduce VOCs. Ideally, the air immediately external to the lab should be conditioned to 18–22°C at 35–55% relative humidity, with a minimum of 15 air changes per hour. The air stream is then processed through activated carbon to remove high molecular weight organics, primarily petrol-derived substances. Potassium permanganate filters can also be used to oxidize lower molecular weight compounds, but should always be used in combination with activated carbon. High efficiency particulate air (HEPA) filters remove only particles smaller than 0.3 μm, not the smaller molecules of pollutants; filters are available to remove VOCs.

Some air purification systems also incorporate UV filters. Parallel streams of HEPA filtered air (laminar flow) blow across the room to expel any dust and particles in the airflow by the shortest route, moving at a uniform velocity of 0.3 to 0.45 meters per second.

Table 8.4 Airborne particulate cleanliness classes (by cubic meter)

ISO class	Number of particles per cubic meter by size in microns					
	0.1 micron	0.2 micron	0.3 micron	0.5 micron	1 micron	5 microns
ISO 1	10	2				
ISO 2	100	24	10	4		
ISO 3	1000	237	102	35	8	
ISO 4	10000	2370	1020	352	83	
ISO 5	100000	23700	10200	3520	832	29
ISO 6	1000000	237000	102000	35200	8320	293
ISO 7				352000	83200	2930
ISO 8				3520000	832000	29300
ISO 9				35200000	8320 000	293000

Box 8.1 Case report

Some IVF clinics operate on a system of "busy" and "quiet" weeks/periods. No cell culture activities are carried out during quiet periods. This type of system can reduce costs of media and personnel, but it does have accompanying risks, as illustrated by this case report. A laboratory with long-standing good reputation, and with a correct and functioning quality management system installed a very efficient and expensive air purification system, due to very poor external air quality associated with industrial coalmining activities. The schedule for servicing and filter replacement was well-defined, with records of new batches of filters and date of replacement. After one particular period of inactivity (quiet period), a new series of cases was initiated, and it was observed that the cytoplasm of all oocytes lysed completely within 3 hours of oocyte collection. Troubleshooting procedures came up with nothing, and the cases had to be transferred to another center willing to take them over. It was initially thought that the expensive air filters had been replaced with a cheaper product, but this was not confirmed by records/orders. Further inquiry revealed that although the filters were correctly replaced, building works had taken place during the quiet period, and some of the air conditioning conducting systems had to be moved over a longer distance, involving an extension in length. The tubing/pipes used for the extensions were of a completely different type, and were coated with a substance that turned out to be responsible for the oocyte lysis. The situation returned to normal when the new parts were removed and replaced with the original type of tubing.

Lessons to be learned:

1. Working with a schedule of quiet and busy periods demands detailed and particular attention, with ongoing system re-validation.
2. Maintaining a small but continuous cell culture activity is easier to monitor than is shutting down with correct start-up each time activity is resumed.
3. No building works should be undertaken without prior announcement to all departments in the hospital. Every change in the building (including painting) that might affect the results of the IVF laboratory should first be discussed with the laboratory director.
4. Maintaining good relations with neighboring clinics is mutually beneficial even if they might be competing for business; in some emergency situations, transferring cases to another clinic may be the only option.

Table 8.5 Types of clean areas (A–D) for aseptic handling and processing of products that are to be used in clinical treatment

Grade	US Federal standard 209E	ISO 14644-1	Description of use
A	Class 100	ISO 5	High risk operations that require complete asepsis, carried out within laminar flow biological safety cabinets (BSC).
B	Class 100	ISO 6	Provides the background environment for a Grade A zone, e.g. clean room in which the BSC is housed.
C	Class 10000	ISO 7	Clean area for carrying out preparatory stages in manufacture of aseptically prepared products, e.g. preparation of solutions to be filtered.
D	Class 100000	ISO 8	Clean area for carrying out less critical stages in manufacture of aseptically prepared products, such as handling of components after washing.

Air flow can be either vertical down-flow (entering via filters in the roof, exiting through vents in the floor) or horizontal cross-flow (entering through filters in one sidewall and exhausted above the floor in the sidewall opposite and/or recirculated via a bank of filters). Air flow of sufficient velocity over critical areas will sweep particles away from the area, and prevent thermal migration of particles from the laminar flow. At least 20 changes of air per hour are usually required for clean rooms classified as Grade B, C, and D (see Table 8.5).

The airflow pressures in clean areas should be continuously monitored and recorded; filters must be regularly inspected and changed to ensure that their efficiency is not reduced.

Guidelines suggested by regulatory and legislative authorities for IVF laboratories are based on those for Good Manufacturing Procedures (GMP), which define four different types of clean areas (A–D) for aseptic handling and processing of products that are to be used in clinical treatment (Table 8.5). Each area has recommendations regarding required facilities, environmental and physical monitoring of viable and non-viable particles, and personnel attire/PPE.

Laminar air flow (LAF) hoods and biological safety cabinets (BSC)

Flow hoods and biological safety cabinets are enclosed workspaces that provide protection either to workers, the products being handled, or both. They can be used to sterilize/clean air entering the cabinet by passage through a HEPA filter and/or the use of UV light or heat.

A cabinet with horizontal laminar air flow (LAF) is a "clean bench," and not a biological safety cabinet. It provides protection only for the product inside, and must not be used when working with any type of chemical or biohazard. HEPA filtered air is directed from the back of the cabinet, across the work surface toward the user and into the room. These should never be used for handling cell cultures or infectious materials, since the operator and the environment will be exposed to aerosols of infectious or allergenic materials. They can be used for preparing culture dishes and tubes prior to use, as well as for aliquoting. Similar "Clean Bench" cabinets with a vertical LAF generally have a sash, and are used in hospital pharmacies to prepare hazardous agents and IV drugs such as chemotherapeutic agents. Air flow originates from the roof of the cabinet, moves

Box 8.2 Case study (*With kind permission from Amy Sparks, University of Iowa*)

An IVF laboratory was equipped with air handling systems that involved 100% fresh air supplied by one handler, and the exhaust managed by another handler. The engineers who installed the system failed to take into account the consequences: it was very inefficient, and the lack of air recirculation made the laboratory vulnerable to toxins in the outside air. A 2-week fire that included burning rubber from >1 000 000 rubber tires posed a tremendous threat, which was managed by doubling the number of potassium permanganate filters. Over a period of 6 months, IVF KPIs started to drop:

- Failed fertilization with standard IVF
- Delayed fertilization, faint PN at 18 hrs post-IVF/ICSI
- Failed cleavage at 40 hrs post-insemination (cleavage rate dropped from 96% to 80%)
- Reduced embryo quality.

The laboratory tried changing every product used, checked hCG criteria, investigated the drug used for hCG, assessed sperm capacitation, etc. Maintenance of the HVAC system was questioned, and the laboratory manager was assured that the filters were up to date. Eventually, she consulted the construction manual herself, and asked another staff member to check all of the filters in place . . . he reported that there were no carbon filters. The installation engineers had removed the carbon filters to remedy a drop in room pressure, without reporting this to the laboratory manager!

This laboratory used large incubators with CO_2 in air, and they were uncertain as to whether the problem might have been avoided had they been using a pre-mixed tri-gas supply. Interestingly, cryopreserved PN zygotes thawed and developed beautifully during this period, suggesting that the period affected by the poor air quality was between oocyte retrieval and fertilization. They ultimately persuaded the hospital to allow them to contact patients who had not achieved an ongoing pregnancy during this period: 90 patients were contacted and offered a free cycle, nearly all of whom accepted the offer.

downwards and is discharged either via a sash into the room, or captured in a vent located on the base of the cabinet.

BSCs provide protection for the product, personnel, and the environment; different types of BSC are designated according to the type and degree of hazard containment required, and the type of protection they provide: Class I, IIa, IIb, and III (see Elder, Baker & Ribes, 2004, Ch. 13, or Elder & Dale, 2010, Chapter 8 for full descriptions). Known (or potentially) infectious agents must always be handled within a BSC, and not an LAF hood. Class II cabinets are generally recommended for use in an IVF laboratory. They have a partially open front, negative pressure, and have a vertical LAF that is HEPA filtered on entry into the cabinet, and during exhaust out of the cabinet. These cabinets provide a microbe-free environment for manipulations, and can be modified to accommodate microscopes and other equipment. However, the presence of equipment can create turbulence and disrupt the LAF, reducing efficiency and safety. Material within the cabinet should be kept to a minimum, with heating stages and light sources preferably built in, and any modifications should be tested and certified to ensure that the system operates correctly. Maintenance and cleaning can be facilitated by the absence of sharp edges and corners.

The LAF can have a significant cooling effect on culture dishes, affecting both temperature control and osmolality through evaporation (see Chapters 4 and 6). Fan noise can be an additional problem. Adjustable bench heights help in carrying out manipulations ergonomically and safely.

Box 8.3 Case report

The EUTCD requires strict adherence to air quality. An IVF laboratory had wrongly interpreted the EUTCD to require that sperm processing must take place in Grade B air, and therefore placed a centrifuge within a laminar flow hood. This constituted a major hazard, as the centrifuge was not on a firm stable base; placing this large piece of equipment inside the flow hood also obscured the air flow and prevented provision of the correct grade of air.

There is a recent tendency to construct new IVF laboratories according to a "Clean Room" design; although this may be a good idea in principle, all parts of the system need to be validated/verified, with incorporated monitoring and elimination of VOCs. Clean rooms are also costly to construct, whereas less expensive options can provide suitable air quality for gamete and embryo processing.

Volatile organic carbons (VOCs)

VOCs in the environment

VOCs are organic (hydrocarbon) compounds that evaporate readily because they have a high vapor pressure at room temperature, i.e. they have low boiling points (usually <100 °C), and many are gases at room temperature. This means that large numbers of molecules from the compound can evaporate or sublimate to enter the surrounding air. VOCs are numerous, varied, and ubiquitous, and include both man-made and naturally occurring compounds released from plants, etc. The "scent" of perfumes, aftershave, lotions, etc. is due to release of VOCs. Some are dangerous to human health, or cause harm to the environment. Anthropogenic VOCs are regulated by law, especially indoors, where they may be concentrated at high levels. Harmful VOCs are typically not acutely toxic, but their effects can be compounded, with a long-term effect on health. They are usually present at a low level, and any symptoms of exposure may be very slow to develop, so that research into VOCs and their effects is difficult.

Many VOCs are chemicals used and produced in the manufacture of paints, adhesives, petroleum products, pharmaceuticals, refrigerants, fuel compounds, solvents, paint-thinners, dry cleaning agents, etc. IVF laboratories located in buildings within polluted areas, or close to airports or industrial manufacturing sites may be exposed to serious VOC contamination that can affect pregnancy and live birth rates. The toxic effects of VOCs have been confirmed in numerous diverse areas, including development of rat and mouse embryos (Hudák & Ungváry, 1978; Shen *et al.*, 2009). A substantial body of published literature now confirms adverse effects of VOCs in an ART setting (Cohen *et al.*, 1997; Boone *et al.*, 1999; Esteves *et al.*, 2004; Lawrence, 2012).

New buildings contribute to the highest level of indoor VOC pollution due to the abundance of new materials generating VOC particles simultaneously over a short period of time. The presence of numerous consumer products makes the total concentration of VOC levels much higher in an indoor environment. Levels during the winter can be 3 – 4× higher than during the summer, when windows are more likely to be open. Sufficient ventilation can reduce VOC levels, but this is not compatible with clean room recommendations for IVF laboratories. In construction of a new laboratory, the appropriate materials must be carefully selected: VOC-free paints and low emission flooring such as VOC-free linoleum.

Box 8.4 Natural sources of VOC

The majority of natural VOCs are produced by plants, mainly isoprene. The rest are produced by animals, microbes, and fungi/molds. The strong odor emitted by some plants is due to a subset of VOCs, green leaf volatiles emitted from the leaves. The rate of biosynthesis is affected by sunlight, and temperature determines rates of VOC volatization/emission. Examples of anthropogenic VOCs are listed below:

1. Paints and protective coatings that require a solvent for spreading; typical solvents include aliphatic hydrocarbons (ethyl acetate), glycol ethers, and acetone. Under pressure of environmental concerns and regulation, many industries are now shifting towards aqueous solvents.
2. Chlorofluorocarbons were formerly widely used in cleaning products and refrigerants; their use is now highly regulated. Tetrachloroethane is used in dry cleaning and in industrial processes. The petrochemical industry produces VOCs both directly and indirectly as by-products.
3. Benzene, a known human carcinogen, is found in tobacco smoke, stored fuels, and exhaust from cars. Natural sources include volcanoes and forest fires. It is also used in the production of plastics, resins, and synthetic fibers. Benzene evaporates quickly, and the vapor is heavier than air, allowing it to sink into low-lying areas.
4. Methylene chloride is found in adhesive removers and aerosol spray paints. In the human body it is converted to carbon monoxide, and will produce symptoms of carbon monoxide poisoning.
5. Perchloroethylene is used in dry cleaning, and is thought to be associated with respiratory problems related to VOC exposure. Studies have shown that low levels can be inhaled from wearing dry-cleaned clothes, and in homes where dry-cleaned materials are stored.
6. Methyl ter-butyl ether (MTBE) is a volatile compound with a minty odor, similar to diethyl ether, used as a gasoline additive to raise the octane number. It was banned from use in the USA in 2004, to limit contamination of water-drinking aquifers from leaking underground storage tanks, but is still widely used in other parts of the world.
7. Formaldehyde is slowly emitted from many building materials, including paints, adhesives, wall boards, and ceiling tiles. Emission is affected by relative humidity, and wood-based materials will release higher levels at high relative humidity and high temperature: a "clean room and VOC-free" environment can easily be polluted by installing furniture made from compressed wood boards containing high levels of formaldehyde.
8. Silicone sealing and tubing are commonly used in building to make weather-resistant joints, which can contain up to 80 g/L VOC. Silicone tubing can emit up to 38 $\mu g/m^3$ acetaldehyde.
9. PVC tubing produces 18 $\mu g/m^3$ acetaldehyde, but when damaged or cut this amount doubles, with additional release of acetone.

Common sources of VOC pollution that can affect an IVF laboratory include:

1. Urban and dense suburban areas: VOCs are produced by industry, vehicle and heating exhausts, and by a variety of cleaning procedures.
2. Substances derived from building materials: aldehydes from flooring adhesives, substituted benzenes, phenol and n-decane released from vinyl floor tiles (flooring adhesives have been found to be particularly aggressive in arresting embryo development). Many paints contain embryotoxic substances, and newly painted surfaces can present a hazard.

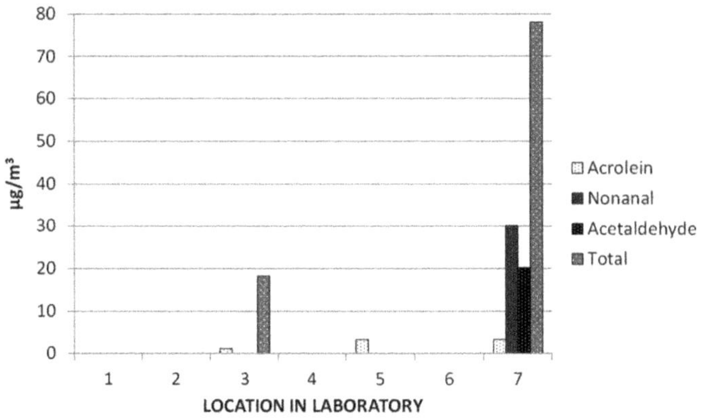

Fig. 8.5 Accumulation of aldehydes in incubators (acrolein, acetaldehyde, and nonanal); the numbers on the *x*-axis indicate the location of incubators in different areas of the laboratory. *(Adapted with permission from: Hall J, Gillian A, Schimmel T, Cecchi M, Cohen J (1998). Hum Reprod,* ***13****(Suppl 4), 146–55).*

3. Compounds released by pesticides, or by aerosols containing butane or iso-butane as propellants. Liquid floor waxes may also contain heavy metals that are embryotoxic.
4. Instruments such as microscopes, monitors, or furniture (as a result of manufacturing processes). A new incubator can have VOC concentrations more than 100-fold higher than used incubators from the same manufacturer: it is important to ensure that all new equipment is allowed to off-gas before use.
5. Perfumes, aftershaves, and other highly scented aerosols – theater and laboratory staff should not wear these, and patients instructed not to wear them when attending for any procedure, from egg collection to embryo transfer.
6. Plastic culture dishes can release VOCs, and should always be off-gassed before use (See Fig. 8.6a). Newly sterilized items continue to release VOCs for prolonged periods of time.
7. Oil used for culture overlays can also act as a "sink" to accumulate organic compounds (see Fig. 8.6b).

The importance of VOCs in the IVF laboratory environment has been comprehensively reviewed by Cohen *et al.* (1997) and Hall *et al.* (1998). Significant levels of benzene, toluene, xylenes, and styrene were found in and around a typical ART laboratory, with the highest concentrations found inside large traditional box incubators (Cohen *et al.*, 1997). Air supplies to these incubators come directly from laboratory ambient air; gas mixtures are supplied in gas bottles, which may be contaminated with organic compounds; a compressed CO_2 gas bottle contains a wide range of VOCs. Anesthetic gases, refrigerants, cleaning agents, hydrocarbons, and aromatic compounds can be detected in an IVF laboratory, and some accumulate specifically in incubators (Hall *et al.*, 1998). Accumulation of VOCs derived from adjacent spaces or specific laboratory products (including sterile Petri dishes) may mean that unfiltered outside air can in some circumstances be cleaner than HEPA filtered laboratory air, or incubator air!

VOC measurement

Detection of VOCs is based on different principles and interactions between the organic compounds and the components of the sensor. Some electronic devices can non-selectively detect ppm concentrations, and others can predict their molecular structure with reasonable

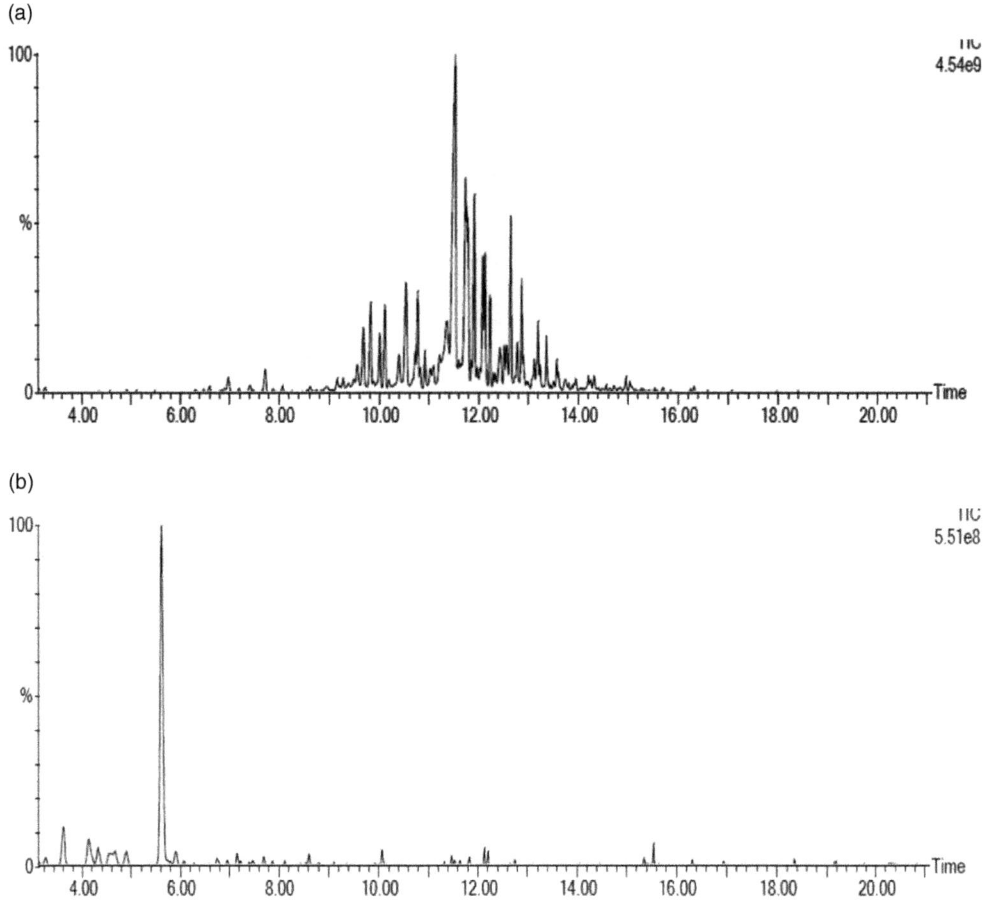

Fig. 8.6 Gas chromatography of passively absorbed VOCs from a polystyrene tube and oil used for culture. (a) Disposable plastic supplies were placed in a VOC-free glass bottle with a gas chromatography assay column, and closed with a VOC-free stopper. The column was sent for analysis after 5 to 7 days' exposure. The figure shows the results of VOCs emitted from a sterile polystyrene tube used for oocyte collection. (b) Gas chromatography of passively absorbed VOCs emitted by oil used as culture overlay. An open bottle of oil was placed in an incubator that had been tested as VOC-free, together with a gas chromatography assay column. The oil showed one major peak, corresponding to methyl-cyclohexane in an amount equivalent to 0.4 ppm Toluol.

accuracy. Solid-phase microextraction (SPME) techniques can collect low concentrations of VOCs for further high-performance liquid chromatographic (HPLC) analysis.

1. *Photo ionization monitors (PID).* These meters are very sensitive and can be used to screen equipment and consumables to pinpoint sources of VOCs, but they do not detect low molecular weight hydrocarbons. An ultraviolet light source is used to break down atmospheric VOCs into positive and negative ions, and the PID detects or measures the charge of the ionized gas as a function of VOC concentration. A PID does not change, burn, or otherwise permanently affect the sample gas – the gas ions recombine to reconstitute the original gas or vapor.

Each VOC has a characteristic photon energy (measured in electron volts, eV) that causes it to fragment into ions when directed at the PID. This is known as the ionization potential

Box 8.5 Case report

A high-performing IVF unit located in a large urban area suddenly experienced a drop in success rates. Troubleshooting did not identify a specific problem: no new methods, new personnel, problems with medium batches. After several weeks of investigation, it was discovered that a nearby supermarket had opened a new extension to its car park. The design of this new car park included evacuation of car exhaust gases, and it happened that the exhaust system was released very close to the laboratory fresh air intake. No smells had been detected in the laboratory, and the air filtration system was properly installed and maintained.

Lessons to be learned:

1. Harmful products such as VOCs do not necessarily have an odor, and human noses are very inefficient at detecting VOCs.
2. The laboratory had no regular survey of air quality that might have detected VOCs.
3. Be vigilant and aware of activities in areas near your clinic: route tarring, roof replacement, paint production.

(IP). PID lamps can be made with a number of different gases, each with a different photon energy, e.g. krypton gas, which has a photon energy of 10.6 eV. The intensity of PID lamps can change over time, or differ from lamp to lamp, and this affects the absolute VOC measurement. PID sensors with the same lamp type will have the same photon energy.

2. *Colorimetric ("Stain") tubes*, or Draeger II tubes have been available for many years, and although they are a proven inexpensive means of measuring toxic gases and vapors, they have limitations. They measure only a single time point, whereas PID sensors can be used for continuous monitoring. They are also slower to respond, so that readings take minutes instead of seconds. The readings are based on color change, which requires human interpretation. The accuracy of the readings also depends on the accuracy of the sampling device, and the colorimetric tubes may or may not have the chemistry required to detect certain compounds.
3. *Flame ionization detectors (FID)* can be very sensitive, are not selective, and can detect all organic compounds, including methane. They are useful in a fixed laboratory site, but cannot be used for portable VOC monitoring due to their size, weight, and need for a hydrogen cylinder.
4. *Thermal desorption or Tedlar®* sampling bags can be used for retrospective analysis of VOCs adsorbed in soil samples, other solids, liquids and gases. The sample is sent to a laboratory for analysis by gas chromatography/mass spectroscopy (GC/MS). This is a good technique for pinpointing a specific problem, but cannot provide real-time detection. The analysis can also be expensive.
5. *Metal oxide semiconductor sensors (MOS)* are compact and inexpensive, but are affected by humidity, and have a non-linear response with long-term drift. They also respond to inorganic gases, and should not be used to assess VOC in the vicinity of higher concentrations of nitric oxide (NO), nitrous oxide (N_2O), or carbon monoxide (CO). MOS can often give both false-positive and false-negative results. MOS sensors may be useful to detect VOCs that are not measured by a PID, although the result may not be very accurate.
6. *Portable gas chromatograph/mass spectrometer (GC/MS)*. Technology traditionally used in analytical chemistry laboratories has been adapted with micromachined silicone

Table 8.6 Highest detected levels of VOC emitted by a Falcon Polystyrene tube 352057

Retention time	Product	CAS Number	Match factor	Concentration (µg/m^3 eq. Toluol)
11.55	Octane, 3-methyl	2216–33–33	86%	7698
11.75	Octane, 3-methyl	2216–33–33	85%	5262
11.91	Octane,25,2,6-trimethyl	62016-28-8	85%	2040
12.09	3-Ethyl-3 –methyl-pentane	17302-01-1	86%	1370

CAS number = unique identification number for a chemical substance, Chemical Abstract Service.
Match factor indicates the degree of correspondence between the detected product and the reference product.

Micro-Electro-Mechanical Systems (MEMS) so that portable systems are available, but the cost is prohibitive. These devices can measure VOCs once every few minutes, but cannot be used as a continuous monitor. It has the advantage of being highly selective, but size, cost, the need for a vacuum pump, and maintenance requirements mean that it should be an alternative only when all else fails.

In the IVF setting, a practical approach is to first assess the presence of VOCs with a portable PID, and then investigate any levels above 0.5 ppm at the ppm level, using passive VOC absorption over several days and further analysis by GC/MS. Very low levels of VOCs can be measured by passive absorption on a gas chromatography column: one side of the column is opened, and left in the room to be analyzed for 5 to 7 days. The column is then closed and sent for analysis. Results are reported as a graph, with peaks as a function of retention time (= the time after which the product is released from the column during chromatography). The height of the peak indicates the amount, and the retention time indicates the type of molecule by comparison with retention times of known chemicals and correspondence calculation. Figure 8.6a and 8.6b are examples of this method used to test disposable plastics and oil used as culture overlay. Table 8.6 indicates the highest detected levels of specific VOCs emitted by a polystyrene test tube used for oocyte collection.

VOC elimination

The principles of VOC elimination/destruction have been described (Khan *et al.*, 2000); a number of processes are available to remove or destroy VOCs on a large scale, but these are not accessible or appropriate in an ART setting. Active filtration units that combine activated carbon filters, catalytic oxidation, and high UV irradiation have been designed specifically for IVF laboratories, and these are highly effective in reducing or eliminating VOCs. Other specialized filters can be installed inside incubators or in-line between the gas cylinder and incubator. Isopropanol and ethanol are commonly used disinfectant agents in the laboratory: they are not only amongst the largest VOC pollutants, but can also absorb material from plastic disposables. Active filtration units can be installed inside an incubator: they filter at a rate of 300 ml/min (18 l/hr), and a single circulation of the entire volume of a 100 l incubator will take 8 hours. An experiment was carried out to test

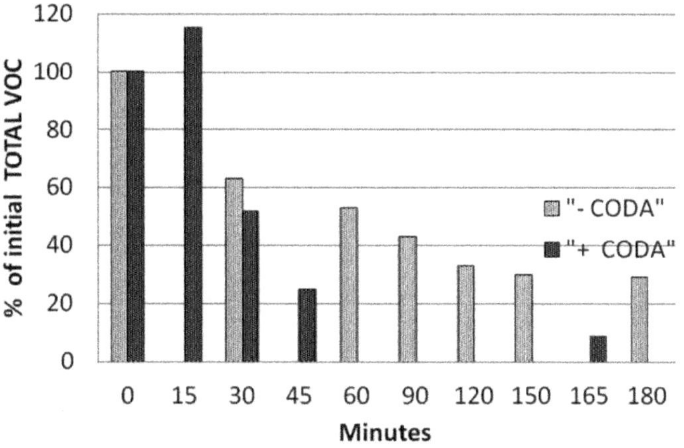

Fig. 8.7 Experiment to test the effectiveness of a "CODA" VOC filter placed inside a 100 l incubator with a Trigas atmosphere after deliberate exposure to isopropanol. The total VOC level was measured every 15 minutes and expressed as a percentage of the initial level measured. VOCs were reduced to 25% after 45 minutes in the presence of the filter, and to less than 10% after 165 minutes. *(Quartec GmbH.)*

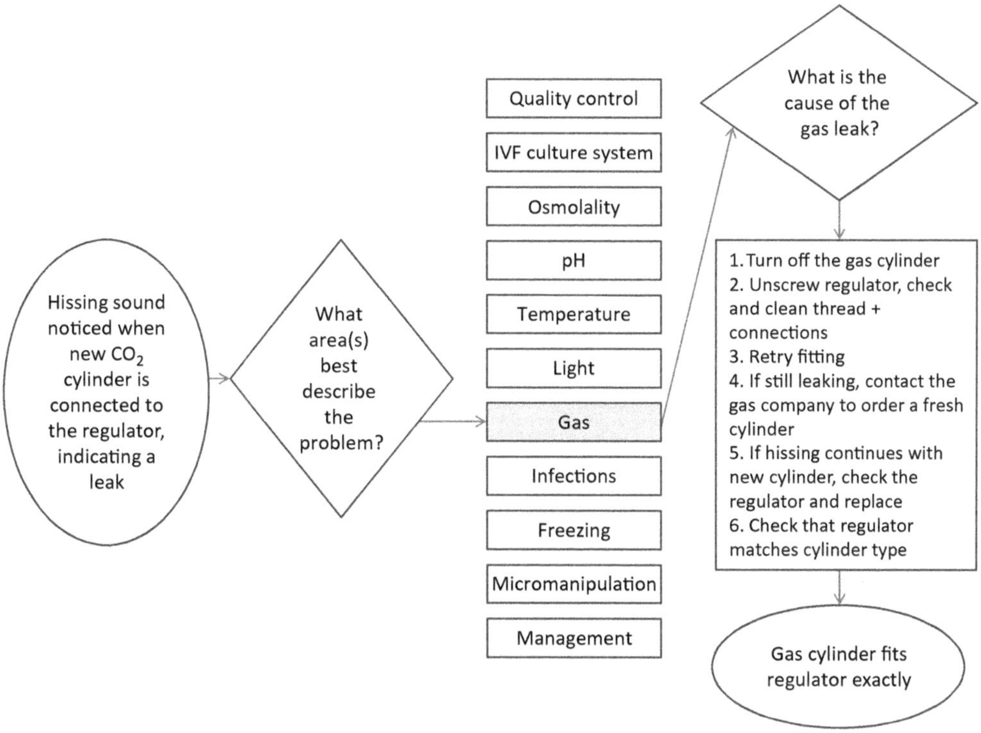

Fig. 8.8 Process map for troubleshooting a gas supply problem.

the potential benefit of using an incubator VOC filtration unit after deliberate introduction of isopropanol (see Fig. 8.7). Hands were disinfected with isopropanol and placed inside a 100 l incubator with a Trigas (5% CO_2/5% O_2/90% N_2) atmosphere for 30 seconds. The door was closed, and VOCs measured via the gas outlet. Without the VOC filtration unit, the demand Trigas atmosphere was restored with a 50% reduction

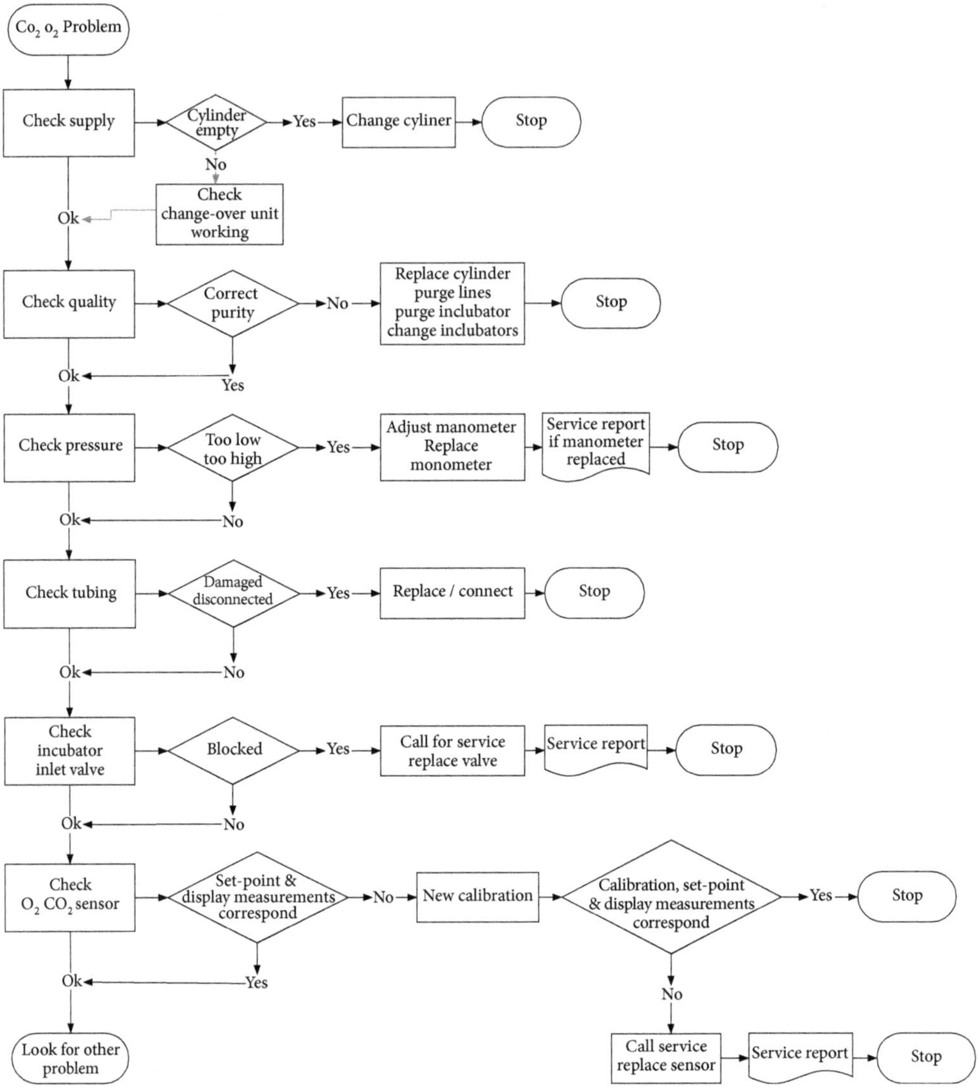

Fig. 8.9 Flowchart for troubleshooting a gas supply problem.

in VOCs after 60 minutes, and 30% of the initial VOC level remained after 180 minutes. With the filter in place, the VOC level was reduced to less than 25% after 60 minutes, and to less than 10% after 165 minutes. Since the large amount of N_2 added by the Trigas mixture effectively reduces atmospheric 20% O_2 to 5% O_2, thereby flushing out VOCs in the incubator atmosphere, VOC reduction in an incubator with an atmosphere of 5% CO_2 in air would be much less effective via the normal gas outlet in the absence of a filtration unit than with the Trigas example above. The efficacy of in-line and extra-line VOC filters has been confirmed in numerous studies; stand-alone filtration units ("towers") can be used to filter air in the laboratory (Schimmel *et al.*, 1997; Esteves *et al.*, 2004; Higdon *et al.*, 2008; http://www.codaair.com/coda-studies.html).

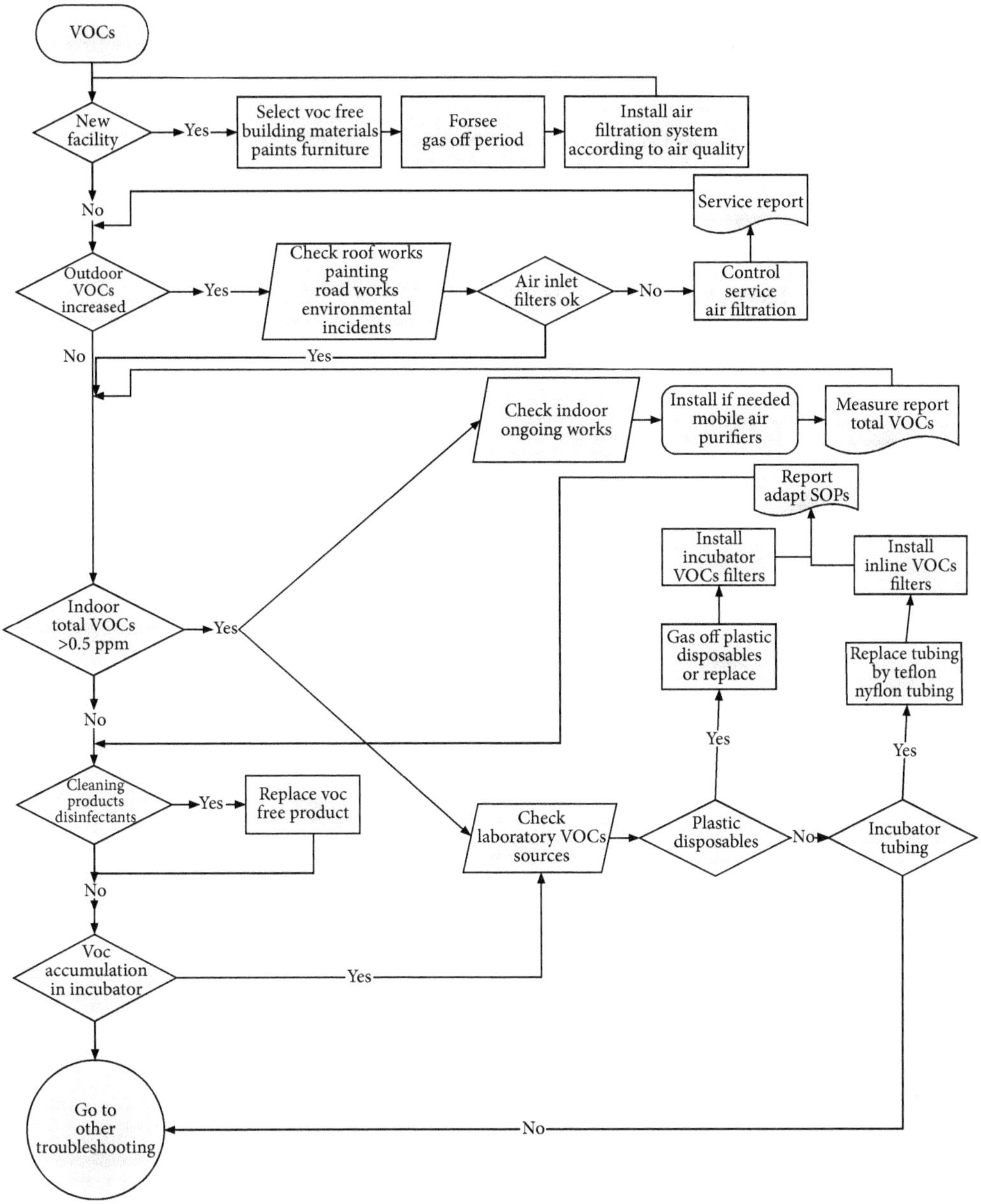

Fig. 8.10 Flowchart for troubleshooting a potential VOC problem.

Conclusions

The potential toxicity of VOCs in ART is now undisputed, and numerous stories continue to circulate, although no specific compounds or molecules have been as yet directly associated with a toxic effect. Direct effects on human embryogenesis have not been described, but these do exist for rat and mouse embryos (Hudák & Ungváry, 1978;

Shen *et al.*, 2009). The first step should always be to remain alert, with an awareness of the problem. Keep watch for any changes in the inside or outside environment that might act as a new source of VOCs (road tarring operations around the clinic are particularly troublesome...), and monitor carefully on a regular basis. Use a stepwise analysis to identify any potential sources of problems, so that strategies can be designed to address any issues, including careful handling of new instruments, equipment, and plastics to minimize release of VOCs into the laboratory environment.

Further reading

Elder K, Baker D, Ribes J (2004). *Infections, Infertility and Assisted Reproduction.* Cambridge University Press.

Elder K, Dale B (2010). *In Vitro Fertilization.* Cambridge University Press.

Guns J, Janssens R, Vercammen M (2012). Air quality management. In *Practical Manual of In Vitro Fertilization, Advanced Methods and Novel Devices*, Nagy, Varghese, & Agarwal (eds) Springer, NY; pp. 17–25.

Mentjes, M (2012). CO_2 and low-O_2 incubators. In *Practical Manual of In Vitro Fertilization, Advanced Methods and Novel Devices*, Nagy, Varghese, & Agarwal (Eds) Springer, NY; pp. 61–70.

References

Boone WR, Johnson JE, Locke AJ, Crane MM, Price TM (1999). Control of air quality in an assisted reproductive technology laboratory. *Fertil Steril*, **71**, 150–4.

Cohen J, Gilligan A, Esposito W, Schimmel T, Dale B (1997). Ambient air and its potential effects on conception *in vitro*. *Hum Reprod*, **12**(8), 1742–9.

Esteves SC, Gomes AP, Verza S (2004). Control of air pollution in assisted reproductive technology laboratory and adjacent areas improves embryo formation, cleavage and pregnancy rates and decreases abortion rate: comparison between a Class 100 (ISO 5) and a Class 1.001 (ISO 6) cleanroom for micromanipulation and embryo culture. *Fertil Steril*, **82**, 349.

Hall J, Gillian A, Schimmel T, Cecchi M, Cohen J (1998). The origin, effects and control of air pollution in laboratories used for human embryo culture. *Hum Reprod*, **13**(Suppl 4), 146–55.

Higdon HL III, Blackhurst DW, Boone WR (2008). Incubator management in an assisted reproductive technology laboratory. *Fertil Steril*, **89**(3), 703–10.

Hudák A, Ungváry G (1978). Embryotoxic effects of benzene and its methyl derivatives: toluene, xylene. *Toxicology*, **11**(1), 55–63.

ISO 6141:2000. Gas analysis – Requirements for certificates for calibration gases and gas mixtures.

ISO 14644-1:1999. Cleanrooms and associated controlled environments; Part 1: Classification of air cleanliness.

Khan FI, Ghoshal AK (2000). Removal of Volatile Organic Compounds from polluted air. *J Loss Prevention in Process Indust*, **13**, 527–45.

Lawrence C, Mortimer S, Havelock J, Mortimer D (2007). VOC levels in a new IVF laboratory with both central and in-laboratory photocatalytic air purification units. *Alpha Newsletter*, No. **36**, 1–5.

Schimmel T, Gilligan A, Garrisi GJ, Esposito Jr. B, Cecchi M, Dale B, Cohen J (1997) Removal of volatile organic compounds from incubators used for gamete and embryo culture. *Fertil Steril*, **68**, Suppl. 1, S165 (Abstract P-151).

Shen S, Yuan L, Zeng S (2009). An effort to test the embryotoxicity of benzene, toluene, xylene, and formaldehyde to murine embryonic stem cells using airborne exposure technique. *Inhal Toxicol*, **21**(12), 973–8.

Chapter 9

Infections in IVF culture

Introduction

IVF culture systems are designed to promote and sustain cell metabolism and optimal cell growth, and therefore also represent an ideal environment for micro-organisms to thrive. Microbial growth has the potential to not only jeopardize the health of gametes and embryos, but also compromise the treatment of other patients via the culture system or nosocomial infection. Although infection/contamination of a culture system is obvious to detect in some cases, it is prudent to bear in mind that undetected "silent" infections can have disastrous consequences, not only in compromising embryo viability, but also in transmitting infection to other patients via their embryos.

All aspects of an ART treatment cycle offer multiple potential sources of microbial contamination, from patients and their biological fluids, fresh and frozen gametes or embryos imported into the lab, staff, visitors, the laboratory environment, air conditioning/heating systems, equipment, consumables, media and imported gases, and liquids (e.g. carbon dioxide, tri-gas, and liquid nitrogen). Micro-organisms of clinical importance to the laboratory include bacteria, fungi, parasites, viruses and, more recently, prions. Numerous publications document cases of infection in ART laboratories (Cottell *et al.*, 1996; Kastrop *et al.*, 2007) and confirm that the threat of microbial contamination is constant and real. Low levels of contamination that are not clinically apparent (e.g. mycoplasma) can also interfere with embryo development by disrupting the balance of media substrates via their metabolism, with release of potential toxins into the medium.

In March 1996, the announcement of a possible link between BSE (Bovine Spongiform Encephalopathy, or "Mad Cow Disease") and a new variant of Creutzfeldt–Jakob disease (CJD) in humans caused a major crisis that devastated the cattle industry in the UK. The infective source of the BSE epizootic was a prion that damages the central nervous system, and was considered to be similar to a prion disease in sheep known as scrapie. This crisis raised concerns about human forms of prion disease and also raised further concerns about the use of blood-derived products as media supplements in ART procedures, prompting increased awareness about the potential hazards associated with handling biological fluids (See Chapter 5 in Elder, Baker & Ribes, 2004).

Blood-borne viruses carried in body fluids can readily be transmitted, and transmission has been documented. In 1991, the use of a serum pool which was positive for Hepatitis B (HBV) resulted in HBV infection of 22 embryos during IVF procedures (Van Os *et al.*, 1991; Quint *et al.*, 1994). During 1998: a patient was infected with HIV following IUI with fresh semen from a donor who had tested seronegative (Matz *et al.*, 1998).

General principles

An active infection control policy is mandatory in every IVF laboratory, with emphasis on asepsis in all aspects of work. The primary objective must be to exclude all possible pathogens that may be present in the sources listed above, and to have an effective policy of containment should infection be introduced (Table 9.1).

Microbes commonly present on human skin (and nostrils) include staphylococci, corynebacteria, *Acinetobacter*, *Klebsiellae*, fecal coliforms and, yeast. Figure 9.1 illustrates that bacterial colonies accumulate in hand crevices and on tips of fingers. Methodical and rigorous hand-washing is mandatory, and hand-washing instruction is an essential training

Table 9.1 Examples of micro-organisms that may be found in patients presenting for ART

Gram-positive bacteria: *Gardnerella vaginalis*, Lactobacilli
Gram-negative bacteria: *Neisseria gonorrhoea*
Treponema pallidum
Chlamydia trachomatis
Ureaplasma/Mycoplasma spp.
Fungi: *Candida* spp.
Parasites: *Trichomonas vaginalis, Schistosoma haematobium*
Viruses: HIV
Hepatitis B (HBV)
Hepatitis C (HCV)
Herpes simplex viruses (HSV)
Cytomegalovirus (CMV)
Human papilloma virus (HPV)

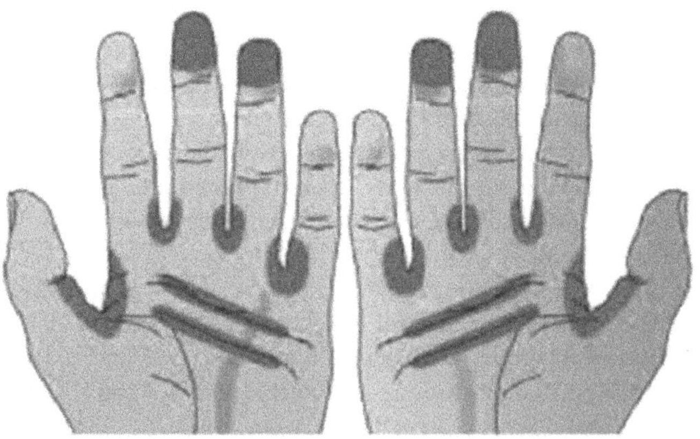

Fig. 9.1 Microbial colonies accumulate in hand crevices and on tips of fingers, illustrated in darkened areas here.

Box 9.1 Case report

A patient had three routine ICSI procedures carried out using frozen–thawed samples of testicular sperm. On each occasion, oocytes were fertilized, but all of the embryos lysed within a few days. The frozen specimen was sent for microbial culture, with a result that showed the presence of gram-negative bacilli, identified as *Morganella morganii;* the bacilli were probably being injected into oocytes along with the sperm. This case also highlights the need to prepare frozen–thawed samples via repeated washing through media containing antibiotics, even when the sperm is of poor quality or obtained from surgical sperm retrieval.

component for all embryologists before they are allowed to handle human gametes or embryos. An unscented antimicrobial skin cleanser (e.g. containing chlorhexidine), can be used, as for pre-operative surgical hand disinfection and for general hospital antiseptic hand-washing.

Anything brought into the lab can be a source of contamination, including consumables, equipment (new or repaired), fresh gametes (e.g. specimen containers for semen samples), frozen gametes, or embryos imported from another lab: this includes infections that may be carried on the outside of dry shippers. Staged entry to the embryology lab is therefore preferable to a single entrance leading to joint embryology and andrology labs.

The wide range of problems associated with microbial contamination in an IVF setting has been comprehensively reviewed (Pomeroy, 2010). An online survey that included 32 experienced laboratory directors reported the incidence and type of contamination found in their IVF laboratories; the results revealed that semen (32%) and inadequate sterile technique (23%) were the most common sources of microbial contamination. The contaminant was identified as bacteria in 49% of the cases, and as yeast or mould (fungi) in 51%. The majority occurred in culture systems that included antibiotics in the media, and therefore the microbes identified (*E. coli, Aspergillus, C. albicans,* gram-negative cocci) had probably acquired resistance to the antibiotics being used. The most common sources of contamination in the cases documented were semen, the technician, and oil.

Normal flora

The human body acts as a natural host to a micro-ecosystem of flora that normally do no harm, unless the natural balance of their habitat is disturbed. Surface tissues in contact with the environment are always colonized with a mixture of microbes, known as "normal flora", and this mixture varies according to site, age, sex, diet, health, and temperature of the environment. The results of a semen or vaginal culture will usually report organisms that are incidental findings of normal flora; antibiotic treatment may disturb the balance, allowing overgrowth of more insidious and resistant pathogens. Normal flora that are benign to the patient can be carried over into an IVF culture system: unlike the human body, embryos have no "defense" mechanism, other than the physical barrier of the zona pellucida, and are highly vulnerable to infection – most cases of overt bacterial contamination will result in non-viable embryos.

Skin is colonized with a variety of normal flora, and shedding skin contributes heavily to the bacterial load of the operating theater and laboratory environment (Dalstrom *et al.*, 2008; Pomeroy, 2010). A poor standard of hygiene, cleaning, and waste disposal will result

Table 9.2 Active spectrum and mode of action of antibiotics commonly used in ART culture media

Antibiotic	Effective against	Mode of action
Penicillin	Gram-positive bacteria	Disrupts cell wall stability
Streptomycin	Gram-positive bacteria	Disrupts protein synthesis
Gentamicin	Gram-positive bacteria Gram-negative bacteria	Disrupts protein synthesis

in a much higher risk of transmitting infection to and from patients, staff, and incubators/embryos. An untidy cluttered lab also increases the risk of contamination; simple cleaning can reduce the environmental microbial load by 80%, and the use of a disinfectant can reduce it to 95% (Pomeroy, 2010). Lab protocols must incorporate a discipline of rigorous hygiene, with

- thorough hand-washing
- use of sterile gloves where appropriate
- regular cleaning of walls, floors, and installations
- autoclaving lab clothing and autoclavable instruments.

Other precautions include HEPA air filtration under positive pressure to ensure that clean air flows into the laboratory, as well as the use of antibiotics in culture media. Table 9.2 shows the active spectrum and mode of action of antibiotics commonly used in culture media. Penicillin has a half-life of 2 hours at 37 °C, and therefore this has now been generally replaced with a stable form of gentamicin. Antibiotic stability is readily affected by temperature, pH, and light, and media must therefore be handled appropriately with this in mind: repeated removal from the refrigerator/opening of bottles and holding media at temperatures >4 °C affect stability of the antibiotics. The following guidelines are recommended:

1. Keep containers tightly closed and refrigerated at all times.
2. After use, discard bottles of culture media that have been warmed.
3. Use media within a few days after opening.
4. Use smaller volume (single-use) bottles whenever possible.

Commercial media are supplied in bottles with a range of different volumes, from 20 ml to 60 ml; the volumes appropriate for individual laboratories will depend upon the daily caseload.

Semen samples

Semen samples are not sterile, even when produced under conditions of strict hygiene. A wide variety of bacteria can be isolated from semen, and the species vary with different country populations. Common normal flora include *Staphylococcus epidermidis*, streptococci (non-hemolytic, ß-, and *Streptococcus viridians*), diptheroids, coliforms, etc. Ureaplasma and mycoplasma species can also be isolated from semen samples (Cottell *et al.*, 1997). More sensitive methods that use PCR to detect bacterial DNA (Kiessling *et al.*, 2008) revealed that 65% of the men tested were found to carry bacteria, including gram-positive anaerobic cocci, *Corynebacterium* spp., Lactobacilli and Streptococci.

A similar study indicated that many of the bacteria in semen are also present in the vaginal flora of some women, suggesting that heterosexual partners may share bacteria (Hou *et al.*, 2013).

Semen may also be contaminated with *Chlamydia trachomatis*, a sexually transmitted infection whose incidence continues to rise each year: 89 million new cases were thought to occur internationally in 2010 (Akande *et al.*, 2010). *C. trachomatis* can bind to sperm and form inclusion bodies (Wolner-Hanssen & Mardh, 1984); its concentration is reduced, but not necessarily removed, during sperm preparation procedures (Al-Mously *et al.*, 2009). Routine clinical diagnosis of chlamydial infection is made by testing a urine specimen or urethral swab, but the organism can be present in semen even if these are negative. Eley & Pacey (2011) reported the results from 11 studies, which showed chlamydial detection in both semen and urine; there was a 20% detection rate in urine only and a 23% detection rate in semen only (Eley & Pacey, 2011).

Mycoplasma species are frequently found in semen, and this microorganism can contaminate a tissue culture without being readily detectable. Ultrastructurally, mycoplasma microorganisms appear as moderately electron-dense spherical particles that adhere to cell membranes and affect the axonemal complex of the sperm mid-piece sections (Sylla *et al.*, 2005). Effects of "silent" mycoplasma contamination include growth inhibition, pH shifts, substrate depletion, induction of oxidative stress, ammonia production, secretion of harmful metabolites, and effects on nucleic acid and protein synthesis with preferential depletion of certain amino acids (Smith, 1960; Macpherson, 1966).

Follicular fluid

Follicular fluid (FF) is not sterile, and some of the micro-organisms identified in FF have been associated with potentially adverse effects on ART treatment (Pelzer *et al.*, 2013). A prospective study assessed the presence of micro-organisms in samples from 262 women: vaginal swabs were taken after vaginal prep by washing, in parallel with FF harvested at the time of oocyte retrieval. All of the vaginal swabs cultured showed evidence of 1–11 different species, with *Lactobacillus* spp., *Bifidobacterium* spp., and *Staphylococcus* spp. the most prevalent. Bacteria were also cultured from 99% of the FFs tested; 71% of the FFs contained bacteria that were also present in vaginal swabs taken from the same patient, and were thus considered to be contaminated. Of the FFs, 29% were colonized with bacteria that were not cultured in the vaginal swab from that patient. Bacteria cultured from FF included Lactobacilli, *Actinomyces* spp., Propionibacteria, *Staphylococcus* spp., *Peptostreptococcus* spp., *Bifidobacterium* spp., *Streptococcus agalactae, E. coli*, and *Enterococcus faecalis*. *Propionibacterium* spp., *Streptococcus* spp., *Actinomyces* spp., *Staphylococcus* spp., and *Bifidobacterium* spp. appeared to be opportunistic pathogens and were related to adverse IVF outcomes. In a separate prospective study, 100% of FF tested showed the presence of micro-organisms after culture (Cottell *et al.*, 1996). Interestingly, the presence of lactobacilli seemed to correlate with positive outcomes, perhaps because lactobacilli produce hydrogen peroxide and lactate, which inhibit the growth of other microbes. A further study in which catheter tips were cultured after ET confirmed that the growth of lactobacilli from the vagina and from the ET catheter was associated with an improved pregnancy rate; the live birth rate was reduced in cases associated with the recovery of *Streptococcus viridans* from the ET catheter (Moore *et al.*, 2000).

Fungal contamination

Yeast infections (*Candida* spp., Fig. 9.2) can be introduced into an IVF culture either from patients during follicular aspiration, or from the skin of laboratory or clinical staff. Fungi such as *Aspergillus* (Fig. 9.3) are frequent airborne contaminants, and may sometimes be traced to a heating or air conditioning system. Culture medium does not contain anti-fungal agents. Fungal infection is known to compromise embryo development, although it

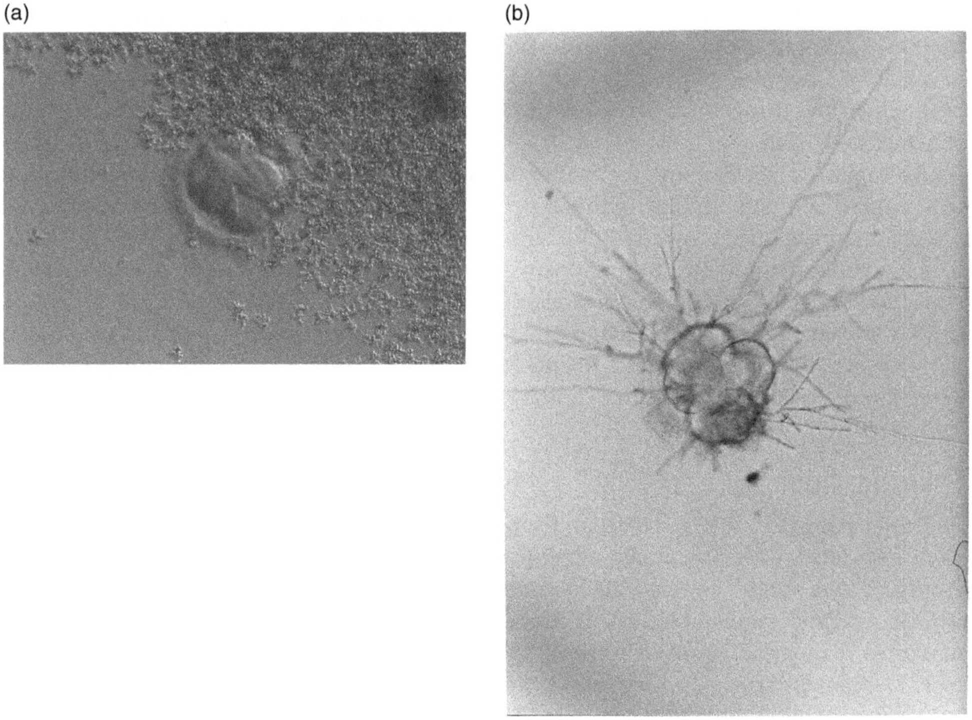

Fig. 9.2 Human embryos with yeast (*C. albicans*) contamination in the culture dish. (a) Embryo with spherical budding cells (blastoconidia) in the medium. (b) Embryo surrounded by filamentous budding elongated cells (pseudohyphae). A black and white version of this figure will appear in some formats. For the color version, please refer to the plate section.

Fig. 9.3 IVF culture dish contaminated with *Aspergillus*.

Box 9.2 Case report

A female patient had been treated for tuberculosis 2 years prior to ART treatment; her partner tested negative for TB. On Day 2 post-OCR, bacteria were identified surrounding the embryos, even though cultures on Days 0 and 1 showed no bacterial growth. The embryos subsequently arrested development. Upon further investigation, it was discovered that the patient had developed a chronic persistent infection with a resistant TB strain, due to previous antibiotic treatment. The conclusion reached was that the bacteria had been transmitted via the egg collection procedure.

Box 9.3 Case report

A couple were treated in the UK for routine IVF treatment. On the day of sample production, eggs from the parasite *Schistosoma haematobium* were identified in a fresh semen sample; the female partner was free of infection. Questioning the couple directly revealed that the man had been walking barefoot in a Malawi lake during a trip earlier in the year. He had acquired the parasite from the waters – *Schistosoma* travels subcutaneously to reside in the bladder/urethra and would have been expelled during ejaculation. All oocytes were vitrified for future insemination, and the patient was prescribed anti-parasitic medication. Subsequent samples were shown to be parasite-free and ICSI treatment was continued on the warmed oocytes (Shearer *et al.*, 2013) (Fig. 9.4).

can be removed from the culture by repeated washing (Ben-Cherit *et al.*, 1996). Embryos that have been exposed to fungal infection within a culture dish can still potentially develop into viable pregnancies, provided that the embryo is immediately and effectively removed from infected medium. Patients must be informed and give their consent before this can be undertaken.

Strategies to minimize infection

A strict policy of avoidance is the best strategy: patient screening via serology for blood-borne viruses, and swabs when indicated by history or clinical examination. All staff must be educated in the principles of aseptic technique and infection control, with an established cleaning rota and SOPs for every procedure. Other routine measures include:

1. Restrict access to the lab, so that only authorized personnel are allowed into the IVF laboratory, wearing specific appropriate laboratory clothing and footwear in the sterile areas of the lab.
2. Thoroughly wash and dry hands and wrists on entering and before leaving the laboratory, and immediately before and after any clinical procedure. Remove watches and bracelets.
3. Hair should be netted in surgical hats. The use of masks is questionable, as the standard surgical face mask has been shown to be ineffective at preventing infection transmission and for use as a respiratory protector. Withdrawal of the face mask in recent years has been on an *ad hoc* basis (Lipp, 2003). If a horizontal laminar air flow is used for media preparation, there may still be some benefit in wearing a mask, to prevent potential transmission of infection from the technician to the sterile preparation area.

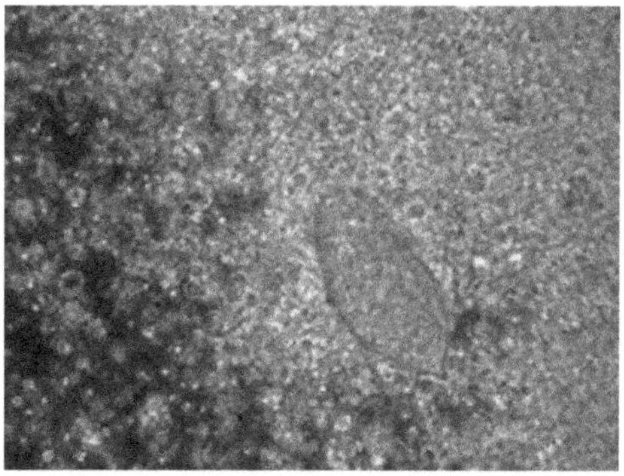

Fig. 9.4 *Schistosoma haematobium* observed in a wet slide preparation during a diagnostic semen analysis (image taken at 400x). (*Photomicrograph by Sue Kenworthy, Lead Andrologist, taken at Portsmouth Hospitals NHS Trust, UK.*) For the color version, please refer to the plate section.

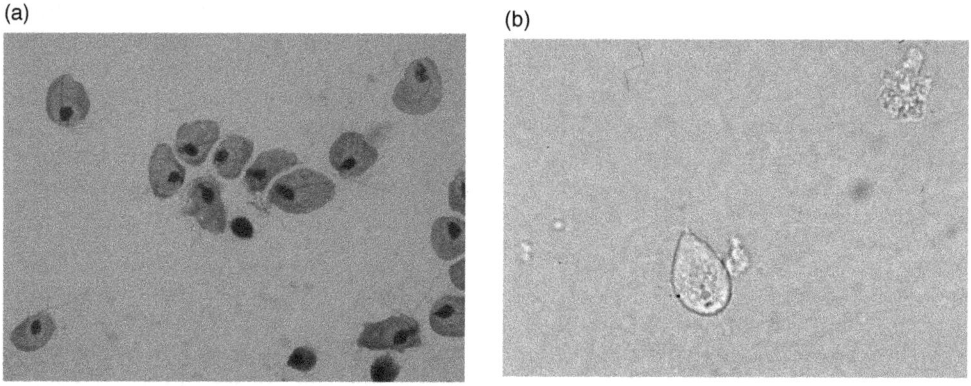

Fig. 9.5 *Trichomonas vaginalis.* (a) Specimen cultured in Diamond's Medium stained with Wright's Giemsa 100x. *T. vaginalis* flagellates are pyriform, round or oval in shape, and measure 7–30 by 5–15 μM. Each organism has a single anterior nucleus and a tuft of four anterior flagella and one short posterior flagellum. A linear axostyle extends posteriorly and projects out of the organism. (b) *T. vaginalis* in a wet mount from urine, unstained 100x, illustrating the pyriform shape and anterior flagellae. A black and white version of this figure will appear in some formats. For the color version, please refer to the plate section. (*Images kindly provided by Julie A Ribes, MD, PhD, Director of Clinical Microbiology, University of Kentucky Hospital, Lexington, KY.*)

Box 9.4 Case report

Trichomonas vaginalis was identified in a sample of surgically retrieved sperm; during oocyte retrieval, the FF was found to be viscous, and cumulus removal was difficult. Only one of 18 oocytes fertilized. Both partners were treated with Metronidazole to clear the infection (Fig. 9.5).

4. A high standard of aseptic technique is mandatory throughout all preparations and manipulations in the IVF lab, in particular when aliquoting media or other liquids and preparing dishes/tubes for preparation and culture.
5. Make sure that plastic sleeves containing dishes/tubes are clean before taking them into the culture environment, and carefully examine all tubes and dishes for any cracks or flaws.

6. Eating, drinking, chewing gum, and applying cosmetics should be strictly forbidden in the embryology lab; smoking is not usually allowed in any clinical area.
7. Keep all work surfaces free of clutter, and keep sleeves of dishes or tubes to a minimum in culture areas.
8. The lab must be easily cleaned and decontaminated. Ideally, the lab design should incorporate curved surfaces between walls, ceilings, and floors, as these are easier to clean and thus less likely to harbor an infection.
9. Separate waste materials according to hazard type (clinical or non-clinical), with sharps always discarded into sharps bins. Clinical waste from patients with a known infection should have separate clinical waste and sharps bins. Ensure that there is an SOP for regular safe disposal of all waste, so that waste is never allowed to accumulate in any sterile area. Remove clinical waste from the lab every day.
10. Handle only a single patient's gametes/embryos at any one time, and use a single aliquot or bottle of media for each patient. Use different micropipettors for dish preparation and insemination procedures. Following incident alerts reported by different UK IVF clinics, the HFEA recommend that micropipettors are allocated solely for these different tasks to prevent cross-contamination. Use a dedicated micropipettor for inseminations only, wiped sterile in between different inseminations. Using long sterile pipette tips where possible will help to make sure that only the tip of the pipette is introduced to the test tube containing the final preparation of sperm.
11. A flowhood dedicated for the sole purpose of dish preparation provides further precaution.
12. Use media with antibiotics that are effective against both gram-positive and gram-negative bacteria.
13. Examine culture media droplets on a regular basis for any evidence of gross bacterial or fungal infection: if any are found, isolate dishes/tubes from those of other patients immediately.
14. Establish a routine for regular cleaning/decontamination of surfaces and equipment. Incubators, flowhoods, centrifuges, and microscopes must be cleaned and serviced at regular intervals.
15. Use safe and appropriate cleaning materials: the agent used must be suitable and effective, but potential toxicity to gametes and embryos that may come into contact with the sterilized material must be considered. Most disinfectants need time to act – they react chemically with living microbes as well as with other materials, so materials should be cleaned first, before being exposed to the disinfectant.

Cleaning protocols

IVF laboratory cleaning protocols must always be carefully evaluated for potentially toxic effects (Hunt, 2008), taking into account the need for off-gassing after cleaning. Investigate the sterilization process used for purchased sterile supplies, and off-gas or rinse with sterile water to ensure that no toxic residue is left. Thoroughly clean all incubators regularly; some have built-in features for cleaning/decontamination, such as heat decontamination cycles (air-jacketed only), hydrogen peroxide sterilization, or ultraviolet treatment for water pans. Any box incubator can be cleaned by removing all of the inner shelves or other interior fittings for cleaning and autoclaving when possible. The interior of the unit can be wiped down with an embryo-safe product, preferably one with low VOC content; if there is an interior fan, take care to keep it clean.

Disinfectants are general biological poisons that kill microbes more effectively than they kill the cells of higher organisms.

1. Chlorine is a good disinfectant, keeps gross microbial contamination under control, and is useful for decontamination after potential exposure to viruses. Commercial disinfectant sprays may be chlorine-based, e.g. Fertisafe™ and Fertisafe Plus™ (Research Instruments, Cornwall, UK).
2. Phenols are good general microbicides, used for floors and walls – but they are powerful poisons, with a strong smell and potential VOC release.
3. Quaternary or cationic detergents are good disinfectants for use on the skin (Oosafe®, Sparmed, Denmark) (Janssens *et al.*, 2007; Lansdowne, 2007).
4. Ordinary soaps and detergents are moderate disinfectants.

Fertisafe™, Fertisafe Plus™ and Oosafe were developed specifically for use in IVF labs. Fertisafe™ contains chlorine dioxide, hypochlorites, ozone, and free oxygen radicals. It does not contain quaternary ammonium or alcohol, has a low chemical load and does not react with nitrogen in biomasses. Fertisafe Plus™ contains an antimicrobial agent, silver dihydrogen citrate (SDC). Oosafe® for hands and incubators contains a quaternary ammonium compound, benzyl-alkyldimethyl chloride 1.3 mg/g (this is not chlorine-based). A second preparation for surface disinfection contains a quaternary ammonium compound (propionate): 6.3 mg/g.

Although some commercial disinfectants claim to have no VOC release and be MEA tested, the details of the MEA method used should be clearly indicated. Catt *et al.* (2013) investigated the effectiveness and safety of three IVF laboratory disinfectants:

1. Commercially available spray (Oosafe®, Sparmed, Denmark).
2. 70% ethanol (short-acting, evaporates quickly, destroys lipid membranes).
3. 6% hydrogen peroxide (destroys proteins and enzymes).

The efficacy of each disinfectant was assessed in three different ways:

1. Surface wiping and swabbing to monitor 48 hr growth on agar plates.
2. 2 μl of each disinfectant was added to culture medium, and this was tested for toxicity using a MEA.
3. Sperm viability toxicity test using medium with disinfectant added.

Their experiments showed that hydrogen peroxide remains the most toxic after cleaning and should be avoided; 70% ethanol is effective, but should be used for incubator cleaning only if a long off-gassing period is incorporated to allow full dispersion of VOC. Commercial disinfectants appeared to be the safest for incubator cleaning, preferably dispensed via a squeezy bottle instead of a spray.

Blood-borne viruses

Blood-borne viruses can remain infectious in dried blood for several weeks:

- HIV and HCV: at least 1 week
- HBV: up to 4 months

Wear protective personal equipment (PPE, gowns, masks, gloves) when dealing with known infectious samples from patients seropositive for blood-borne viruses. Handle samples in a lab/area designated for infectious agents, with dedicated equipment; cases can be isolated in a self-contained independent culture chamber (MIC-101, Billups-Rothenberg, Inc., Del Mar, California – see Chapter 6), scheduled as "last on the list," or during a specified time period when no seronegative patients undergo treatment. Although sperm washing can be effective at reducing and often eliminating the virus from samples obtained from a seropositive man, the process is not 100% efficient (Semprini *et al.*, 1998; Obradors *et al.*, 2014). Therefore precautions are necessary even when handling processed samples.

Liquid nitrogen dewar sterilization

Dewars that have been used for holding quarantined or viral positive patient samples can be disinfected.

- Safely empty the contents of the dewar, and, using standard precautions for handling liquid nitrogen (see Chapter 10), decant the contents in an outside area that is a safe distance from any buildings/cars. Bear in mind there may be loose straws/vials at the bottom if the dewar has been in use for some time and has not been recently emptied.
- Allow the dewar to air dry, and soak the empty dewar completely in a strong detergent, fully wiping the interior.
- Discard the disinfectant and repeat the process.
- Allow the dewar to stand empty for a minimum of 4 months (the longest survival period for HBV). The inside of the dewar can be swabbed for testing to ensure that there is no viral residue.

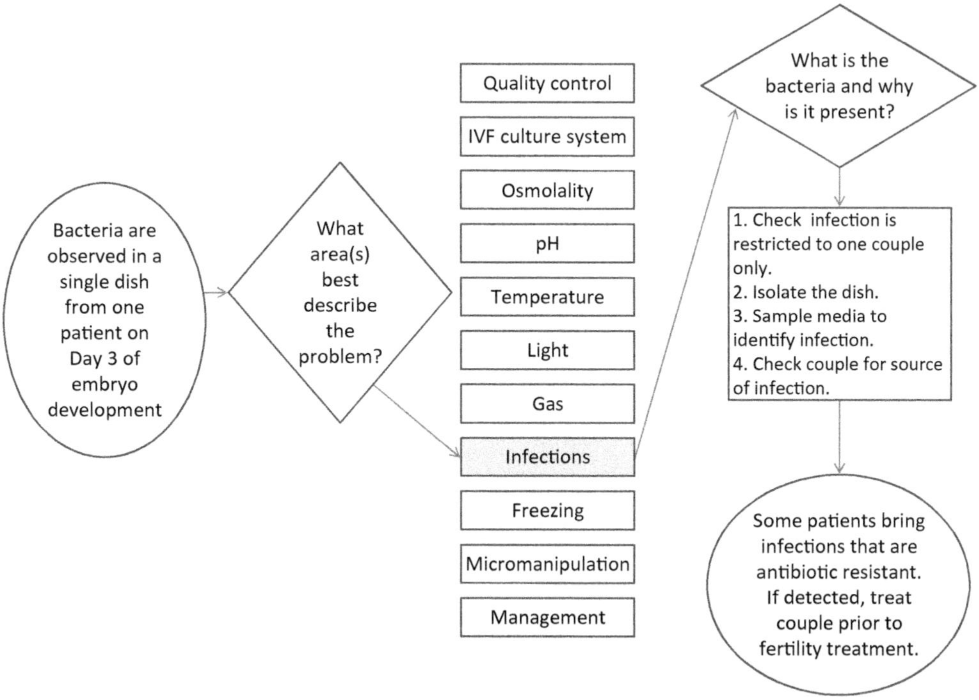

Fig. 9.6 Process map for troubleshooting infections.

Fig. 9.7 Flowchart for troubleshooting infections.

- If tests confirm that the dewar is negative for the known viruses, it can then be used to store screened samples when re-filled with new liquid nitrogen. The exterior of the dewar must be thoroughly cleaned/sterilized before reintroduction into the cryostore.

Over recent years, liquid nitrogen that has been sterilized via exposure to ultraviolet light as a further precaution against cross-contamination has been introduced. The expense of this procedure may prohibit its use, especially since despite much concern, no contamination via liquid nitrogen in the field of ART has yet been reported. However, experimental results show that there may still be a role for the use of sterile liquid nitrogen in vitrification procedures (Parmegiani *et al.*, 2012). Contamination was introduced into two batches of liquid nitrogen, one with bacteria and one with fungi. Several different vitrification devices (Cryotop, Cryoleaf, Cryopette) were immersed in the contaminated nitrogen. "Control" devices were then compared with those "washed" three times in certified ultraviolet sterile liquid nitrogen. Microbiological assessment showed a high level of contamination in the "no-wash" control group: 78.6% of the devices were bacteria contaminated, and all were fungi contaminated. Following three washes with sterile LN_2, none of the devices were found to be contaminated with either bacteria or fungi.

Conclusion

All human tissue should be considered as potentially infectious. Asepsis in all aspects of clinical and laboratory work is essential in order to exclude potential pathogens from every potential source, and prevent transmission of any pathogens that might be introduced. A stringent Infection Control Policy is mandatory in every IVF lab, with protocols that specify aspects of hygiene, labeling, protective clothing, laundry, safe handling of sharps and specimens, disposal of clinical waste, and decontamination procedures.

Further reading

Elder KT, Baker D, Ribes J (2004). *Infections, Infertility and Assisted Reproduction.* Cambridge University Press, UK.

Pomeroy KO (2010). Contamination of human IVF cultures by microorganisms: a review. *J Clin Embryol*, **13**(4), 11–30.

Steyaert SR, Leroux-Roels GG, Dhont M (2000). Infections in IVF: review and guidelines. *Hum Reprod Update*, **6**(5), 432–41.

References

Al-Mously N, Cross NA, Eley A, Pacey AA (2009). Real-time polymerase chain reaction shows that density centrifugation does not always remove *Chlamydia trachomatis* from human semen. *Fertil Steril*, **92**, 1601–15.

Ben-Cherit A, Shen O, Haran E, Brooks B, Geva-Eldar T, Margalioth EJ (1996). Transfer of embryos from yeast-colonized dishes. *Fertil Steril*, **66**, 335–7.

Catt S, Lingham E, Lee W, *et al.* (2013). A randomized trial investigating the effectiveness and safety of three IVF laboratory disinfectants. ESHRE 2013 conference, London, UK.

Cottell E, McMorrow J, Lennon B, Fawsy M, Cafferkey M, Harrison RF (1996). Microbial contamination in an in vitro fertilization-embryo transfer system. *Fertil Steril*, **66**, 776–80.

Cottell E, Lennon B, McMorrow J *et al.* (1997). Processing of semen in antibiotic-rich culture medium to minimize microbial presence during *in vitro* fertilization. *Fertil Steril*, **67**, 98–103.

Dalstrom DJ, Venkatarayappa I, Manternach AL, Palcic MS, Heyse BA, Prayson MJ (2008). Time-dependent contamination of opened sterile operating-room trays. *J Bone Joint Surg*, **90**, 1022–5.

Eley A, Pacey AA (2011). The value of testing semen for *Chlamydia trachomatis* in men of infertile couples. *Int J Androl*, **34**(5 Pt 1): 391–401.

Hou D, Zhou X, Zhong X, *et al.* (2013). Microbiota of the seminal fluid from healthy and infertile men. *Fertil Steril*, **100**(5), 1261–9.

Hunt P (2008). Lab disinfectant harms mouse fertility. *Nature*, **453**, 964.

Janssens R, Cortvrindt R, Devroey P, Lauwers S, van der Elst J (2007). Novel sterilizing solution Fertisafe™ is effective and non-embryotoxic. *Hum Reprod*, **22**(Suppl 1), i221, p563.

Kastrop PMM, de Graaf-Miltenberg LAM, Gutknecht DR, Weima SM (2007). Microbial contamination of embryo cultures in an ART laboratory: sources and management. *Hum Reprod*, **22**, 2243–8.

Kiessling AA, Desmarais BM, Yin H-Z, Loverde J, Eyre RC (2008). Detection and identification of bacterial DNA in semen. *Fertil Steril* **90**(5), 1744–56.

Lansdowne D (2007). Evaluation of a decontamination strategy for a CO_2 incubator using Fertisafe and Sterile Water. Galaxy R CO_2 Incubator. Study commissioned by Research Instruments Ltd.

Lipp A (2003). The effectiveness of surgical face masks: what the literature shows. *Nursing Times*, **6**(99), 22–4.

Macpherson I (1966). Mycoplasmas in tissue culture. *J Cell Sci*, **1**, 145–68.

Matz B, Kupfer B, Ko Y, *et al.* (1998) HIV-1 infection by artificial insemination. *Lancet*, **351**, 728.

Moore DE, Soules MR, Klein NA, *et al.* (2000). Bacteria in the transfer catheter tip influence the live-birth rate after *in vitro* fertilization. *Fertil Steril*, **74**, 1118–24.

Obradors A, Woodward BJ, Mataro D, Vesnaeve V, Vassena R (2014). Semen residual viral load and ART outcomes in 248 HIV Positive men. *Hum Fertil*, 1–5, in press; doi:14647273.2014.932967, Posted online 20th June 2014.

Parmegiani L, Accorsi A, Bernardi S, Arnone A, Cognigni GE, Filicori M (2012). A reliable procedure for decontamination before thawing of human specimens cryostored in liquid nitrogen: three washes with sterile liquid nitrogen (SLN2). *Fertil Steril*, **98**(4), 870–5.

Pelzer ES, Allan JA, Waterhouse MA, *et al.* (2013). Microorganisms within human follicular fluid: effects on IVF. *PLoS One* **8**(3), e59062.

Pomeroy KO (2010). Contamination of human IVF cultures by microorganisms: a review. *J Clin Embryol*, **13**(4), 11–30.

Quint WG, Fetter WP, van Os HC, Heitjink RA (1994). Absence of hepatitis B virus (HBV) DNA in children born after exposure of their mothers to HBV during *in vitro* fertilisation. *Clin Microbiol*, **32**(4), 1099–100.

Semprini AE, Persico T, Thiers V, *et al.* (1998). Absence of hepatitis C virus and detection of hepatitis G virus/GB virus C RNA sequences in the semen of infected men. *J Infect Dis*, **177**, 848–54.

Shearer C, Burchill A, Drezet C, *et al.* (2013). A case study reporting *Schistosoma haematobium miracidia* detected in semen and the outcome on treatment decision. *Hum Fertil*, **16**(3), 223.

Smith PF (1960). Amino acid metabolism of PPLO. *Ann NY Acad Sci*, **79**, 543–50.

Steyaert SR, Leroux-Roels GG, Dhont M (2000). Infections in IVF: review and guidelines. *Hum Reprod Update*, **6**(5), 432–41.

Sylla L, Stradaioli G, Manuali E, *et al.* (2005). The effect of *Mycoplasma mycoides* spp. mycoides LC of bovine origin on *in vitro* fertilizing ability of bull spermatozoa and embryo development. *Anim Reprod Sci*, **85**(1–2), 81–93.

Van Os HC, Drogendijk AC, Fetter WPF *et al.* (1991). The influence of contamination of culture medium with hepatitis B virus on the outcome of *in vitro* fertilization pregnancies. *Am J Obstet Gynecol*, **165**, 152–9.

Wolner-Hanssen P, Mardh PA (1984). *In vitro* tests of the adherence of *Chlamydia trachomatis* to human spermatozoa. *Fertil Steril*, **42**, 102–7.

Chapter 10

Troubleshooting a cryopreservation system

Introduction

Cryopreservation of gametes, embryos, and reproductive tissue is an integral part of an ART service. Options and protocols continue to evolve: successful use of new protocols means that cryopreservation of oocytes, sperm, surgically retrieved sperm/testicular tissue, and embryos at all stages of pre-implantation development is now routine. Many laboratories now also offer cryopreservation of ovarian tissue for specific clinical indications. In common with all other aspects of ART, troubleshooting in cryopreservation requires an understanding of basic principles of cryobiology. This topic has been covered in detail in numerous texts (see Elder & Dale, 2010, Ch. 12; De Santis & Coticchio, 2011; Leibo & Pool, 2011), and only a brief outline of major points is presented here.

Cells exposed to a reduction in temperature below the normal body temperature are subjected to two different classes of physical stress:

1. Direct effect of reduced temperature: changes in membrane properties (permeability and fluidity), premature exocytosis of cortical granules and zona hardening, changes to cytoskeletal structures (e.g. depolymerization of the cytoskeleton and spindle microtubules), change in gas solubility, increased cytoplasmic viscosity and pH.
2. Physical intracellular changes associated with ice formation during slow freezing: water is sequestered, and the residual aqueous phase becomes increasingly concentrated, with dramatic increases in salt concentrations. Cryoprotectant agents (CPAs) are used to act as "anti-freeze compounds" and compensate for these effects, increasing the volume of the residual unfrozen phase and reducing its ionic composition.

Factors that affect cellular response to freezing

1. *Surface area:* oocytes have larger surface area/volume ratios, as well as different intracellular properties, compared to cleavage stage embryos. All embryos have a different surface area/volume ratio compared to sperm. Oocytes and embryos will therefore take longer than sperm cells to reach osmotic equilibrium in solutions containing CPAs.
2. *Cell volume changes:* abrupt changes in cell volume can cause immediate damage and make cells more susceptible to stress during cooling/thawing. Equilibration must be carried out so as to minimize extreme fluctuations in cell volume, whilst at the same time avoiding excessive exposure to the potentially toxic chemicals.

3. *Cooling to subzero temperature:* the optimum cooling rate depends on the type of cell, as well as type and concentration of CPA. The optimal warming rate also depends on CPA type and concentration, and other biophysical factors including cell volume/surface area, permeability to water, and Arrhenius activation energy (Arrhenius' equation accurately describes temperature dependence of reactions). Cooling and warming rates that are optimal for one cell type may be harmful to others.
4. *Osmotic events*: as cells are frozen, they must respond osmotically to large changes in extracellular fluid concentrations. The efflux of water during slow freezing results in osmotic dehydration and causes oocytes and blastomeres to shrink. If cooling rates are too slow, cells are exposed to hypertonic conditions for too long; a rate that is too fast causes the formation of intracellular ice, which is lethal. When external and internal osmolality are in equilibrium, the cell will recover approximately its original volume. The time taken to recover the initial volume is a good indication of how fast a molecule can enter the cell, and also indicates the magnitude of osmotic stress inflicted on the cell. Zygotes are a good model, because the diameter of the cytoplasm can be measured easily, and the percentage loss or regain of volume can be calculated. The minimum equilibration time required in zygote slow freezing protocols can be estimated by plotting the percentage of initial cytoplasmic diameter as a function of CPA exposure time. Time zero = 100%; when a plateau is reached, the cell is in equilibrium with CPA.
5. *The temperature and duration of exposure of the cells to the CPAs:* this may differ at 37 ° C compared with a cooler room temperature of 18–22 °C, and will also differ according to the type and concentration of the CPA.
6. *Removal of the CPA:* when frozen cells are warmed rapidly, the concentrated CPA is rapidly diluted. A rapid influx of water can cause osmotic shock, which can be reduced by using a non-permeating substance such as sucrose.
7. *Cell cytoplasm is converted from liquid to solid state regardless of the freezing method:* a slow frozen oocyte/embryo is essentially vitrified intracellularly, with ice present in the extracellular portion. Intracellular ice is always detrimental, and must be avoided.
8. *Blastocoel collapse:* When blastocysts are vitrified, the blastocoel cavity should initially collapse when exposed to CPAs. This can be initiated artificially, e.g. by administering a single laser pulse directly to the trophectoderm (at a point furthest away from the inner cell mass) or by physical manipulation, via gentle aspiration into a micropipette with a slightly smaller lumen than the diameter of the blastocyst, or via micromanipulation using an ICSI pipette to pierce the blastocoel (Vanderzwalmen *et al.*, 2002; Mukaida *et al.*, 2006; Feng *et al.*, 2010; Iwayama *et al.*, 2011; Min *et al.*, 2013).

Slow freezing of oocytes and embryos

Slow freezing cryopreservation protocols have two basic components:

1. Cryoprotectant solutions that buffer the cell cytoplasm from intracellular damage.
2. Controlled freezing rate and controlled initiation of ice formation ("seeding") at the optimal temperature to ensure cell viability after thawing.

Cryoprotectants

CPAs stabilize intracellular proteins, reduce or eliminate lethal intracellular ice formation, and moderate the impact of concentrated intra- and extracellular electrolytes. All CPAs are hyper-osmotic and mix readily with water. They promote hydrogen ion bonding and thereby reduce the availability of water within the cells to be cryopreserved. CPAs also prevent crystal formation by "trapping" water molecules in between CPA molecules. CPAs are generally toxic compounds, and the level of toxicity varies according to the cryoprotectant, its concentration, the duration of exposure, and the temperature at which CPA exposure takes place. Results from studies comparing relative toxicity among CPAs are inconsistent, and there is no overall consensus regarding CPAs with the highest/lowest toxicity. The situation is further compounded by the use of various combinations of CPAs (see Bernal *et al.*, 2014). Glycerol is a naturally occurring cryoprotectant that has low toxicity (used by the Canadian Wood frog and Japanese salamander during hibernation at temperatures below −50 °C), but it penetrates cells very slowly. CPAs are divided into two groups:

1. *Permeating (intracellular):* penetrate the cell membrane (more slowly than water), stabilize intracellular proteins, reduce the temperature at which ice forms, and minimize osmotic damage resulting from effects of electrolyte concentration. Examples include:
   - 1,2 propanediol (PROH)
   - Dimethylsulfoxide (DMSO)
   - Glycerol (Gly)
   - Ethylene glycol (EG)
   - Butylene glycol (BG)
   - Propylene glycol (PG)
   - Methanol (Met)
2. *Non-permeating (extracellular):* do not penetrate cell membranes. Sugars and polymers contain both hydrophilic and hydrophobic/lipophilic groups (i.e. they are amphipathic compounds), and act as osmolytes, increasing the osmolarity of the extracellular space so that water diffuses out of the cell and ice crystal formation is reduced. They act as osmotic buffers, stabilize cell membranes, and protect against cell swelling during the addition/removal of cryoprotectants.

   Sugars: sucrose, trehalose, D-mannitol, raffinose, lactose.
   Polymers: polyvinylpyrrolidine (PVP), ficoll (Fic), dextran (Dex), polyethylene glycol (PEG)

   Glycine, proline, and trehalose are amphipathic compounds that interact with membrane lipids and proteins to alter phase transitions and hydration state.

   Commercially supplied freeze and thaw media are designed to optimize embryo survival, and may have unique CPA concentrations specific to that media. Media from different companies may contain different buffers, possibly a different protein source, and there may be minor differences in salt concentrations (see Bernal *et al.*, 2014 for list). Companies recommend that freezing and thawing should be carried out using the specific media supplied by the same company. However, if media from the same company is not available and another must be substituted, the latter should have the same CPA that was used for freezing, at a similar concentration. The other differences in buffers/protein/salts will be negligible in the hypertonic thawing solutions. More recently, Parmegiani *et al.* (2014) have proposed that a "universal warming procedure" using a single CPA may be an efficient method for warming any reproductive cell (see p. 159 for further details).

Seeding

Nucleation and continued ice crystal formation is initiated by "seeding", induced ice formation in the sample via direct contact with a liquid nitrogen (LN_2)-cooled implement (e.g. a cotton bud dipped in LN_2). A correct "seeding" process is essential to subsequent sample survival: the difference between laboratories that achieve good results and those that are less successful can often be traced back to this practical step. Some practitioners prefer to seed both ends of an embryo straw to prevent possible embryo movement within the straw as ice crystals form.

Straws can be frozen either horizontally or vertically; embryos sink in the cryoprotectant solution, and will rest at the wall of a horizontal straw, or at the bottom of the liquid column in a vertical straw. For vertical freezing, seeding can be initiated at the surface of the fluid column, i.e. the point furthest away from the position of the embryo. In some vertical freezers, the diameter of the straw holder needs to be considered: if the diameter of the holder is significantly wider than the straw, an additional empty straw can be added to the outside of the straw. This "pushes" the straw closer to the inner portion of the freezing device, and maximizes the effects of controlled temperature drop.

Samples must be "held" at the nucleation temperature, close to −7 °C, for 5–10 minutes in order to ensure that thermal equilibrium is reached; seeding can then be carried out at the level of the meniscus using a cotton bud that has been dipped in LN_2. A spatula or forceps dipped into the LN_2 can be used, but these are less effective – a cotton bud holds more liquid and thus cools more effectively. The local cold spot created leads to ice nucleation, followed by propagation of the ice front through the sample. Confirmation that ice crystal formation has commenced before continuing with a slow freezing program is critical: failure to do so can result in poor freezing technique and poor survival. As ice forms, release of latent heat results in a slight rise in temperature; maintaining the seeding temperature for a further 5–10 minutes during ice propagation allows dissipation of latent heat before freezing is continued. Subsequent slow cooling then leads to increasing cellular dehydration, with water continuing to leave the cell until around −20 °C.

If the cooling rate over a large temperature drop is too fast, the cell(s) cannot dehydrate, and intracellular ice formation is inevitable. The cell remains in a very unstable state down to −30 °C, when a sudden transition from liquid to solid (ice) will occur in microseconds. Since the cell has not been sufficiently dehydrated, lethal ice crystals are formed. Over a period of microseconds, the temperature will first rise from −30 °C to -7 °C (nucleation temperature) and then return to the freezing chamber temperature of −30 °C, after the phase transition from liquid to solid. This represents a considerable thermal shock for the cells. This "undercooling" phenomenon is often poorly understood, so that the importance of correct seeding is overlooked. Seeding in cryovials is much more difficult than seeding in straws, and needs practice before the process can be accomplished effectively.

Cryopreservation protocols

Vitrification (non-equilibrium/ultra-rapid freezing)

Vitrification avoids the formation of ice, combining high concentrations of CPAs with ultra-rapid cooling so that water molecules do not have time to form ice crystals. The cells are therefore not "frozen", but "vitrified" (glass-like, from Latin "*vitrum*"): the samples are rapidly cooled by direct contact with LN_2, and reach a low temperature in a state that has the molecular structure of a viscous liquid without crystals, forming an amorphous glassy solid. The high concentrations of CPAs required can be toxic, and there needs to be a practical balance between maximal cooling rate and lowest effective level of CPAs. Exposure time is important for the cells to reach adequate permeation/dehydration. An initial cooling rate that is greater than the critical cooling rate of the solution is needed to avoid formation of intracellular ice: in human ART, this should be >20 000 °C/min, and preferably >50 000 °C/min. Rapid warming prevents devitrification and ice crystal formation during the transition state. The CPAs are diluted in several steps, in order to counterbalance osmotic effects as the CPAs leave the cell.

Reducing the volume of the sample to a minimum when placing embryos onto the vitrification device is thought to reduce potential CPA toxicity and osmotic damage; current vitrification protocols suggest a volume of 1–2 µl per sample, although the volume can be decreased even further to leave only a "film" around the embryo. However, individual clinics should assess the optimal volume that provides the best survival rate in their systems. Removal of almost all of the fluid prior to plunging into LN_2 can result in poorer survival rates in some circumstances, and leaving at least 1 µl around the embryos may be preferable (BW personal communication).

Theoretically, several oocytes or embryos can be cryopreserved in a single straw or on a single vitrification carrier. Guidelines for good practice suggested by the Association of Clinical Embryologists in the UK recommend that no more than two embryos should be loaded per device (Brison *et al.*, 2012). Loading only one embryo per device presents less potential risk to the embryo(s). Single storage also means that one oocyte/embryo at a time can be thawed/warmed, which optimizes the use of cryopreserved material. Two embryos from a single patient cryopreserved on two devices also allows them to be stored in separate dewars, reducing the risk in case of a dewar failure.

Two types of "carrier" are currently in use:

1. "Open": the sample comes into direct contact with LN_2, cooling rate approx. 23 000 °C/min
   - Open-pulled straw (OPS)
   - Cryoloop
   - Cryoleaf
   - Vitriplug
2. "Closed": the carrier, but not the sample itself, is in direct contact with LN_2, and the cooling rate is slower, approx. 12 000 °C/min
   - Cryotip (Irvine Scientific)
   - CryoLock (Biotech Inc)
   - CryoBioSytem HSV straw
   - Rapid-i, RapidStraw (Vitrolife)

Practical tips for vitrification

1. Pre-equilibrate all media to the correct temperature prior to dispensing, and invert the vials immediately before use to make sure that the solutions are fully mixed.
2. Label each vitrification device fully prior to use. Remember that timing is crucial to effective vitrification and warming, and make sure that everything is "ready to go" prior to starting the procedure.

3. Prepare a LN_2 bath and place it next to the flowhood, ready for plunging the loaded vitrification devices. The vessel should be a properly functioning container dedicated for the purpose of holding LN_2. Portable insulated containers ("eskies") used for media transport/deliveries are not designed for use as LN_2 baths, and using them for this purpose is highly inadvisable. Any leak of LN_2 could have disastrous consequences. Place the LN_2 bath on a flat secure surface; do not use a chair or stool!
4. Carry out all vitrification/warming procedures in an area that will remain free of distraction throughout.
5. Adhere to the correct timings strictly, using a timer. Using two separate timers can prevent the loss of vital seconds in re-setting a single timer. It is also helpful to have a second embryologist to assist with the timings, etc.
6. If a straw sealer is used, make sure that it is switched on and ready to use before starting the vitrification process.
7. Some media companies recommend using 1–2 ml solution volumes for each vitrification/warming event, but 100–200 μl droplets are equally effective, provided that the oocyte/embryo is full equilibrated in each solution. Larger volumes for warming have the advantage that the vitrification device can be easily submerged into a large volume, making removal of the oocyte/embryo easier. Using larger volumes of the initial equilibration solutions also means that they can be warmed for up to 30 minutes without major pH or osmolality disturbance.
8. Ensure that there are no air bubbles on the surface of the initial warming solutions, as embryos tend to adhere to bubbles; this will hinder full submersion into the solution, and also affect the precise timing of the warming process.
9. The speed of warming (around 20000 °C/min) is considered to be more important in avoiding lysis than is the cooling speed. If warming is too slow, the intracellular CPA concentration is too low to prevent ice from re-crystallizing, and the supercooled liquid forms lethal ice crystals.
10. Embryo survival may not be immediately obvious after warming, and survival/morphology is routinely assessed after a minimum period of 2 hours in culture. This is particularly important in the case of blastocysts, which may appear collapsed immediately after warming but will re-expand within 2 hours when cultured under optimal conditions.

Bear in mind that, unlike slow freezing protocols, vitrification has a steep learning curve. Good survival rates reported in the scientific literature take time (and patience) to achieve. The vitrification technique can be practiced using unfertilized oocytes or abnormally fertilized embryos (both with patient consent) before it is applied in treatment cycles. An immediate improvement in results may not be obvious when switching from slow freezing to vitrification; a return to slow-freezing is acceptable as long as success rates are good. However, higher survival rates after vitrification have been demonstrated worldwide (Edgar & Gook, 2012), and perseverance is recommended. Participation in workshops provided by companies that supply vitrification media can be helpful, where experts can demonstrate optimal use of specific devices and media. Visiting a lab that has a good vitrification program can also be helpful.

Equilibrium freezing (slow freezing)

Cells are equilibrated in an aqueous solution with concentrations of 1.0 to 1.5 M CPA and 0.1 M sucrose, and frozen in straws. The temperature is lowered at a controlled rate, and ice crystal formation is initiated by seeding at −6.5 to −7.0°C. Slow cooling is usually continued to −40°C, and then the cooling rate is increased as the temperature is rapidly lowered to −150°C before final transfer to LN_2 for storage. Thawing is carried out by initial rapid warming, e.g. by plunging the straw into a test tube containing sterile water at 30 °C or 37 °C for 30 to 60 seconds. This is followed by serial dilution of the CPA at room temperature, before returning the cells to culture at 37°C.

Practical tips

1. Make sure that no bubbles are trapped within the freezing medium after the samples are loaded into straws. If the column of fluid is unintentionally broken during loading a straw, it is better to return an oocyte/embryo to the fluid and reload. A new straw may be necessary. Air bubbles can sometimes be seen on thawing, and these are a hazard to the fragile dehydrated oocyte/embryo. Air bubbles or small leaks in straws can cause them to explode on thawing (see below). Straws may be placed in a contained box during the initial thawing stage, in case the plug explodes out of the straw. Use of personal protective equipment (PPE), particularly eye shields is strongly advised when thawing straws.
2. Straws have a large surface area, small diameter, and thin wall: they warm very rapidly when removed from a cold environment. If they are held outside the freezing machine for too long during seeding, they can warm to a temperature that is too high for ice nucleation. Local cooling induced by the cold tool may cause local ice formation, but the bulk temperature of the fluid in the rest of the straw may not allow the ice crystal to propagate through the sample. Remove straws only partially to allow visual confirmation that an ice column has formed; removing samples fully from the freezing machine in order to check that seeding has occurred can cause the nucleated ice to melt.
3. Thermal control of the freezing apparatus must be accurate and stable at the seeding temperature. The temperature may be low enough to allow nucleation because of the thermal mass of the cold tool, but not low enough to allow subsequent ice propagation. Any thermal fluctuations within the controlled-rate freezer can also cause the ice to melt.
4. Nucleation of ice in straws at temperatures very close to the melting point results in very slow propagation of ice through the sample. In some cases, ice propagation can become blocked, so that embryos are then effectively supercooled. In this case the embryos would not be expected to survive further cooling.

Warming rates are faster than critical cooling rates (Fahy *et al.*, 1987), and a higher CPA concentration in warming solutions compared to cooling solutions prevents crystallization (Vanderzwalmen *et al.*, 2002). This has led to the proposal that the same warming protocol can be used for both vitrified and slow-frozen oocytes (Parmegiani *et al.*, 2014). Since commercial vitrification warming solutions have an extracellular CPA concentration of 1 M/l (first warming step) and 0.5 M/l (second warming step), theoretically these solutions might be used to warm any reproductive cell, regardless of the freezing protocol (Parmegiani *et al.*, 2014). Using a single warming procedure, irrespective of the freezing protocol/composition or brand of the freezing solution, would simplify freezing activity within embryology labs and could potentially optimize costs in the laboratory. A prospective study designed to test this hypothesis demonstrated that the blastocyst formation rate was significantly improved when 359 slow-frozen human pronucleate

Box 10.1 Liquid nitrogen expansion in the gas phase

Nitrogen changes from the liquid phase to the gas phase with an expansion rate of 682 times – i.e. 1 ml of liquid nitrogen will generate 682 ml of nitrogen gas. This represents a particular hazard for scientists who store samples in vials, as vials routinely leak nitrogen when submerged in liquid. The screw thread is not an effective seal, and for this reason companies that produce cryovials state that they should not be used for clinical storage. When vials are removed from storage, the liquid nitrogen cannot "leak" out, and converts to nitrogen gas at the warmer air temperature. A leaking vial containing 1 ml of liquid nitrogen will inevitably explode with the massive conversion to 682 ml gas.

A properly sealed straw should not leak nitrogen; if there is a tiny gap, liquid nitrogen can enter a straw, causing it to explode on thawing/warming. A survey of embryologists in 2008 revealed that the majority had observed explosions of vials or straws during thawing (Tomlinson & Morroll, 2008). Wearing personal protective equipment (PPE) is therefore essential during every thawing/warming procedure.

embryos were thawed with vitrification warming media instead of a conventional thawing protocol (Kojima *et al.*, 2012).

Controlled-rate freezers

Slow-cooling protocols need temperature reduction that is controlled in an accurate and reproducible manner. Controlled rate freezing machines can be programmed according to specified protocols. Three basic types of controlled-rate freezers have been used in ART:

1. Linear freezers (Planer, CryoLogic)

   The CryoLogic system is free-standing and modular, with a temperature controller, cryochamber, and cryobath. The chamber stands directly in LN_2 in a cryobath, and is connected to the controller which regulates the temperature of the specimens. No special installation is required: systems can be quickly set up and packed away by the user.
2. Non-linear (liquid nitrogen-free, asymptote, Morris, 2006)
3. Progressive (whole ovaries, Arav *et al.*, 2005)

It is important to be aware of any alerts issued by companies that supply controlled rate freezers. For example, Planer issued an alert in 2014 regarding failure of the chamber temperature probe, due to a short circuit. This was a concern in certain specified models of the Planer Kryo freezer models. If the short circuit occurred, the temperature appeared to be too cold, and the controller attempted to compensate by warming the samples. As a corrective action, two software changes were released to improve the resilience of the freezer systems to temperature probe faults. Keeping up to date with such communications is essential in an IVF laboratory.

Cryostore management

The capacity of a cryostore depends on the freezing strategy of the clinic and its activity, i.e. the number of cycles with potential freezing performed per year. The number of dewars/vapor stores can then be predicted, always allowing for a spare dewar for emergency use. Vapor freezers may have a higher capacity, but they are more expensive, and concern regarding temperature stability has been expressed (BW personal communication). Vapor freezers may be preferred for an andrology cryostore, but liquid nitrogen dewars are preferable for vitrified embryos, as immersion in liquid nitrogen prevents temperature variation.

Dewars

Dewars are constructed with an inner and an outer shell separated by a vacuum, which provides superinsulation to minimize temperature loss. All dewars must have written records of service inspections, and pressurized vessels should have a data plate fixed to the outside detailing the Written Scheme of Examination (WSE). They must be inspected regularly for damage, making sure that the vacuum is functional. Visual inspection of the outside can sometimes reveal "cold spots," which may indicate a fault with the vacuum. Significant changes in temperature can be detected by running a hand over the outside of the vessel. Necks and foam/cork lids are particularly prone to damage, which will jeopardize the effectiveness of the dewar. Ice plugs that are potentially explosive can also build up around the neck – this is also the case for vacuum flasks used to decant LN_2. Figure 10.1 illustrates examples of a damaged dewar lid core (a) and neck (b, c).

1. Failure to check and top up dewars can have disastrous consequences for cryopreserved material, and top-up can be managed via a staff rota that is strictly adhered to. Top up only one dewar at a time, checking and recording LN_2 levels prior to filling so that rates of LN_2 loss can be calculated for each container. These records will give an indication of

(a)

(b)

(c)

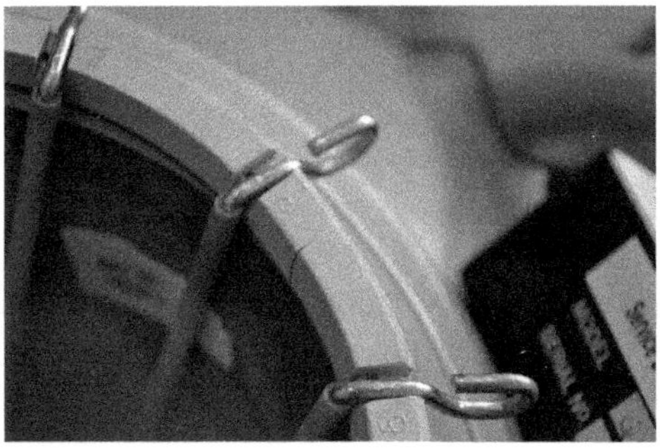

Fig. 10.1 Damaged dewar storage tanks. (a) Two dewar lid cores. The core on the left continued to be used after breaking in half, with significant loss of nitrogen. A replacement lid (on the right) shows the extent of the damage to the core. The dewar maintained nitrogen far more efficiently after replacing the damaged core. (b), (c). Two dewars with obvious signs of damage. Apart from inadequate nitrogen retention, the vessels must have received a significant impact for such damage to be visible: there may have been further damage that was not visible.

Table 10.1 Grade 5.0 liquid nitrogen specification *(courtesy of Mr. Nick Brown, Air Products)*

	Specification	Typical analysis (by volume)
Nitrogen	> 99.999%	> 99.999%
Oxygen	< 5 ppm	< 2 ppm
Water	< 2 ppm	< 1 ppm
Carbon dioxide		<0.5 ppm
Carbon monoxide		< 1 ppm
Nitrous oxide		< 0.1 ppm
Total hydrocarbons		<1 ppm
Methane		< 0.5 ppm
Acetylene		<0.05 ppm

which dewars maintain nitrogen less efficiently, and whether any should be replaced. Filling hoses may be fitted with a handle that avoids handling the hose directly, even with protective cryo-gloves. When not in use, the hose can be kept in a holster fitted to the wall of the cryoroom; the sinter (brass sprinkler at the end of the hose) must be inspected regularly for damage.

2. Dewars should be equipped with an alarm that is triggered by low LN_2 levels, either via a platform base that registers weight change, or with a system that monitors temperature inside the dewar (http://www.networkedrobotics.com).
3. Vapour cryostores require vacuum-isolated pipework to deliver the LN_2; piped LN_2 may be preferable to collecting into large dewars that have to be wheeled from a delivery vehicle (which will inevitably be dirty from exposure to the outside environment) to the cryostore. Pipework for LN_2 can be expensive, and the distance between the cryotank and the hose at the end of the pipework is also important: shorter distances are more efficient, as less LN_2 is lost in transit. The system can be permanently primed by use of a hot bypass valve, but this can be an expensive option. Vapor cryostores are replenished via an autofill system, and this must be inspected regularly to make sure that no blockage occurs. Problems can occur with solenoid blockage and faulty sensors or controllers. The system should also be linked to an emergency power supply, as a loss of power will result in failure to maintain LN_2 levels in the system.

Safety precautions for handling liquid nitrogen

Cryostore specifications

1. *Location:* LN_2 must only be handled in a well-ventilated area, ideally in a room with an external wall and exit door to the outside, or an internal room with fitted external vents. A window in the internal wall/door allows personnel in the room with LN_2 to be visible at all times. Optimally, the room could be designed to be at a slightly lower level (6 inches to 1 foot) relative to adjacent rooms; N_2 gas is heavier than air, and sinks when

LN_2 is decanted. If too much gas is accidentally produced, a lower depth to the cryostore allows access to higher external ground in an emergency, away from the asphyxiating gas. Proper use of a cryostore should ensure that a low O_2 atmosphere is never created.

2. *Flooring:* ideally extending up the wall to a height of 3 inches. Vinyl will crack when exposed to LN_2, and should not be used as a floor covering. A steel plate or thermocrete scree can be layered over the area where LN_2 will be decanted.
3. *Low O_2 monitor with visual and audible alarm:* fixed at a height of 5 feet above the floor **inside** the cryostore room. The alarm usually has two stages, triggered at 19.5% and then at 18.5% oxygen; it should be linked to a digital display repeater outside the room, so that it is absolutely clear whether or not it is safe to enter the room in an emergency. *Do not enter* if the low O_2 monitor indicates that it is not safe to do so, unless specific breathing apparatus is available! Portable low O_2 monitors may be used, but only as an extra safety device, and never in place of a fixed monitor. If you enter a low O_2 atmosphere with only a portable monitor, the alarm will be triggered, but you may immediately collapse due to asphyxiation.

Safety

1. Write an SOP (in layman's terms) to cover correct procedures and behavior for all staff with access to the cryostore and LN_2, including clear instructions regarding emergency procedures and contacts. Basic emergency procedures can be displayed on the wall of the cryostore where they are clearly visible.
2. Training: all staff who work with LN_2 or who might be involved in an emergency must have adequate training, carried out on-site in the cryostore facility by an instructor with comprehensive knowledge of health and safety during LN_2 handling. A written record should document completed training, and no-one should be allowed to handle LN_2 without full training.
3. On-call system for cryostore emergencies out of normal working hours: ideally, the cause of the alarm can be first assessed by the on-call staff in order to decide whether a visit to the clinic is necessary. Immediate attendance is required if there is a danger that cryopreserved material might be compromised. Only staff who are able to get to the clinic within a specified period of time in order to effect a remedy should be included in the on-call rota (See Chapter 12).
4. Risk assessment can determine the likelihood of a cryostore incident, and the potential severity of outcome. Not all risks can be accounted for (e.g. natural disasters – see Chapter 12), but all avoidable risks should be considered. Risk can be minimized by reducing the amount of LN_2 stored, and by regular checks of all monitors and extraction systems. Low level O_2 and N_2 alarms can be tested weekly when dewars are topped up. Practice drills can ensure that everyone knows what to do in an emergency. If pressurized vessels or dewars need to be wheeled through a public area, e.g. a public corridor of a hospital, first carry out a risk assessment of the route. If elevator transport is required, use it only for the LN_2 vessel, with no accompanying personnel: a leak could cause asphyxiation. Block the elevator for LN_2 use only, with exits on all floors blocked off by the clinic team. One member of the team can send the vessel in the elevator to the required floor, and a receiving member of the team can remove the vessel at its destination.

(a)

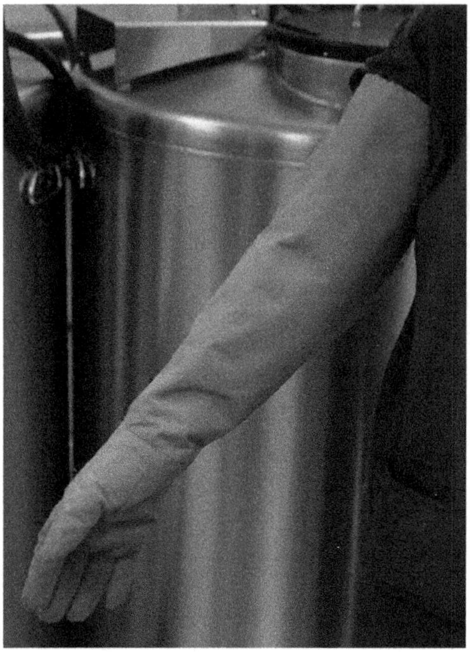

(b)

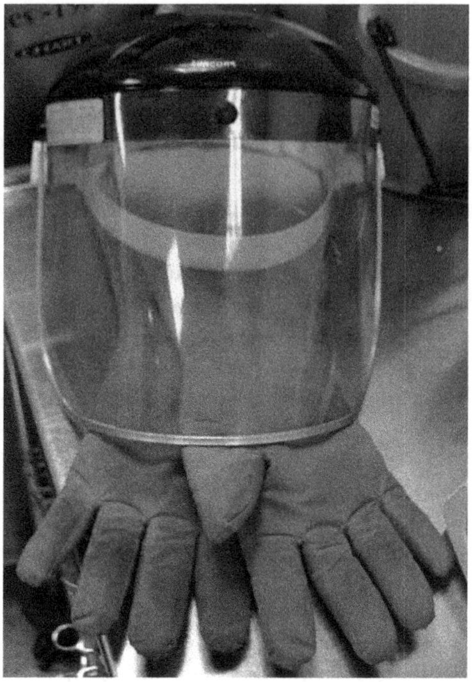

Fig. 10.2 Personal protective equipment used for liquid N_2 handling. (a) Long cryoglove. (b) Full face guard and cryogloves. (*Images kindly provided by the Hewitt Fertility Centre, Liverpool Women's Hospital, UK.*)

5. N_2, as liquid or as vapor, can cause cryogenic burns that must be treated immediately. Never handle LN_2 without personal protective equipment (PPE):
   - A face shield (safety goggles will only protect the eyes, not the face)
   - Dry insulated gloves that should be loose fitting for quick removal in an emergency, but ideally with an elasticated wrist band to prevent nitrogen from splashing into them (Fig. 10.2)
   - Do not wear boots or clogs unless trousers are fully tucked **outside** them – any LN_2 that splashes into a boot/clog cannot escape, and will cause burns
   - An overall or apron with long sleeves is also advisable.

If LN_2 has splashed onto clothing, remove it immediately if possible, bearing in mind that removing "frozen" clothing directly stuck to the skin can cause more skin damage. Cotton clothing can 'hold' LN_2 for several seconds, and this will increase the risk of burns if it is in contact with skin. If cotton socks are worn in cryogenic boots, always tuck trousers completely outside them. Be aware that extremely low temperatures can cause materials to become brittle, so that they break easily. Skin that has been in contact with a cryogenic agent should be gently immersed in cold water for a minimum of 10 minutes. Do not use warm or hot water, or dry heat. Depending on the severity of the burn, return of capillary blood supply will gradually turn the skin color red. Cover the affected areas with a sterile dressing, and request medical assistance.

Box 10.2 Case study

Two young researchers were found unconscious next to three LN_2 dewars in a University cold room, and died in hospital 1.5 hrs after cardiopulmonary resuscitation. Dark red discoloration on the skin of the left arm was found on one of the researchers at autopsy. Histology confirmed that the skin damage was inflicted by cold due to direct LN_2 exposure before death. The cause of death was asphyxia due to O_2 deficiency, and the skin histology demonstrated LN_2 contact (Tabata *et al.*, 1995).

(a) (b)

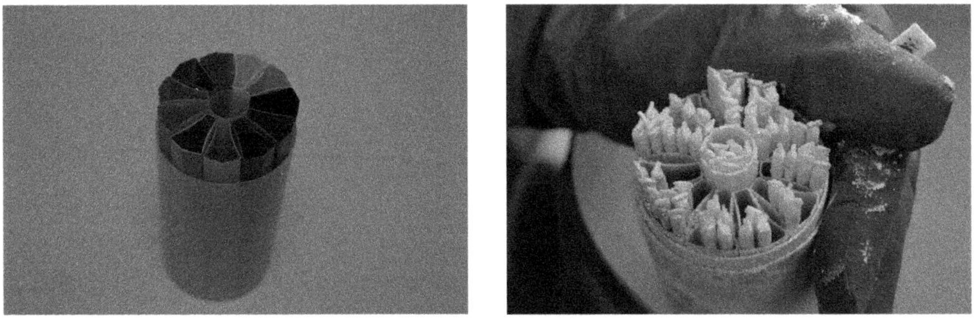

Fig. 10.3 Cryostore management. (a) Colored goblets help in keeping track of samples. They can be arranged in alphabetical order, as shown here: aqua, black, blue, brown, green, gray, orange, pink, purple, tangerine, yellow, or in the order of the rainbow/white light spectrum. (b) A canister that has just been removed from liquid nitrogen, containing straws that are distributed between different colored goblets. A black and white version of this figure will appear in some formats. For the color version, please refer to the plate section. (*Images kindly provided by the Hewitt Fertility Centre, Liverpool Women's Hospital, UK.*)

Cryostore audit

Each canister in a dewar can hold multiple goblets, and quick access is facilitated by using a unique colour for each goblet within the canister (Fig. 10.3). This speeds up placement and removal from cryostorage, and also helps in tracking samples during an audit. Using digits for the individual canes can also be useful. For example, embryos for patient X in "dewar 2," "canister 3," "cane 6," "blue goblet," can be more easily located compared to using a system without number or colors. Needless to say, an accurate up-to-date record of the location of all cryopreserved material is essential.

It is important to maintain contact with patients who have their gametes/embryos in storage, and confirm that storage is in accordance with their wishes. The regulatory bodies of many countries produce their own consent forms to cover a defined period of storage. In-house consent forms can act as legal contracts between the patients and the clinic, clearly defining the action to be taken if the patients fail to maintain contact with the clinic or cannot be contacted.

If a patient's circumstances or terms of consent change, the clinic must take appropriate action, and in a timely fashion. Any communication from a patient to the clinic regarding the terms of storage (i.e. moving overseas, extending storage period, etc.) must be relayed to the IVF laboratory so that that no material is removed from the cryostore during an audit in accordance with older consent forms. A specific email group within the clinic for patients

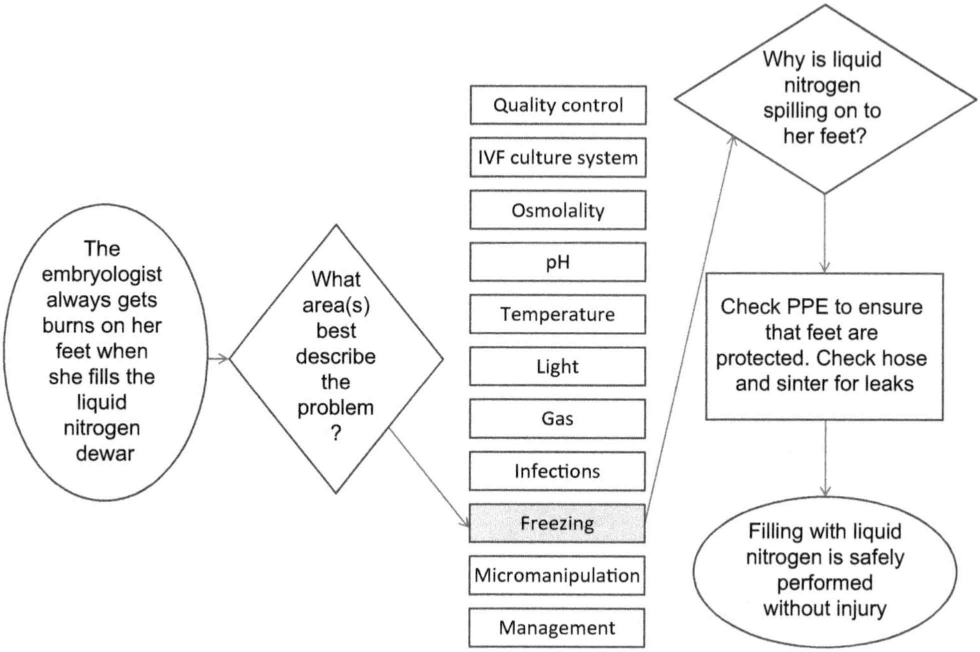

Fig. 10.4 Process map for troubleshooting a LN_2 handling problem.

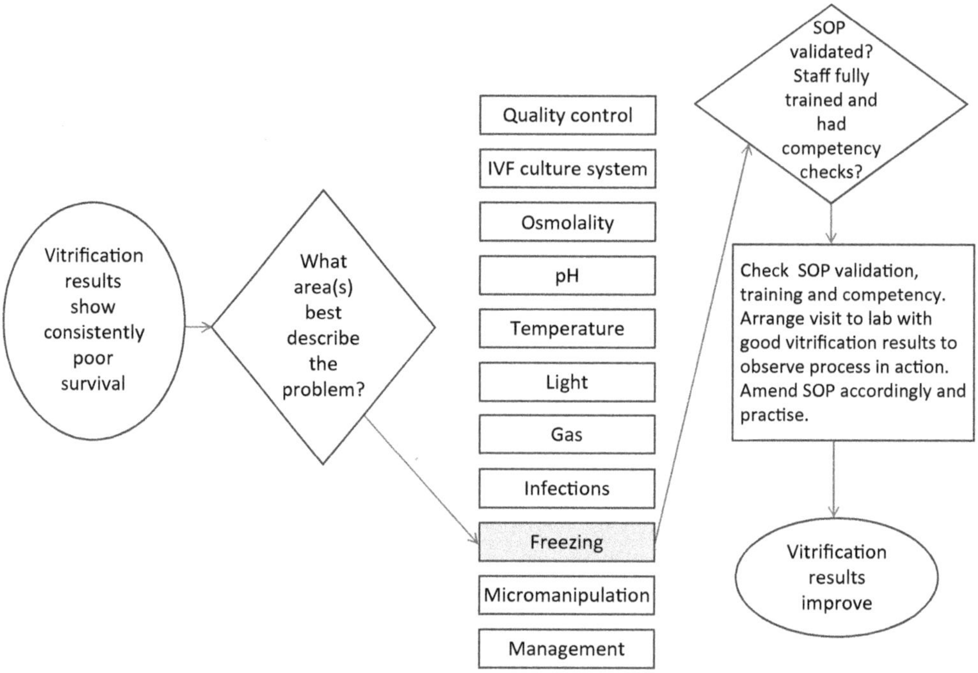

Fig. 10.5 Process map for vitrification troubleshooting.

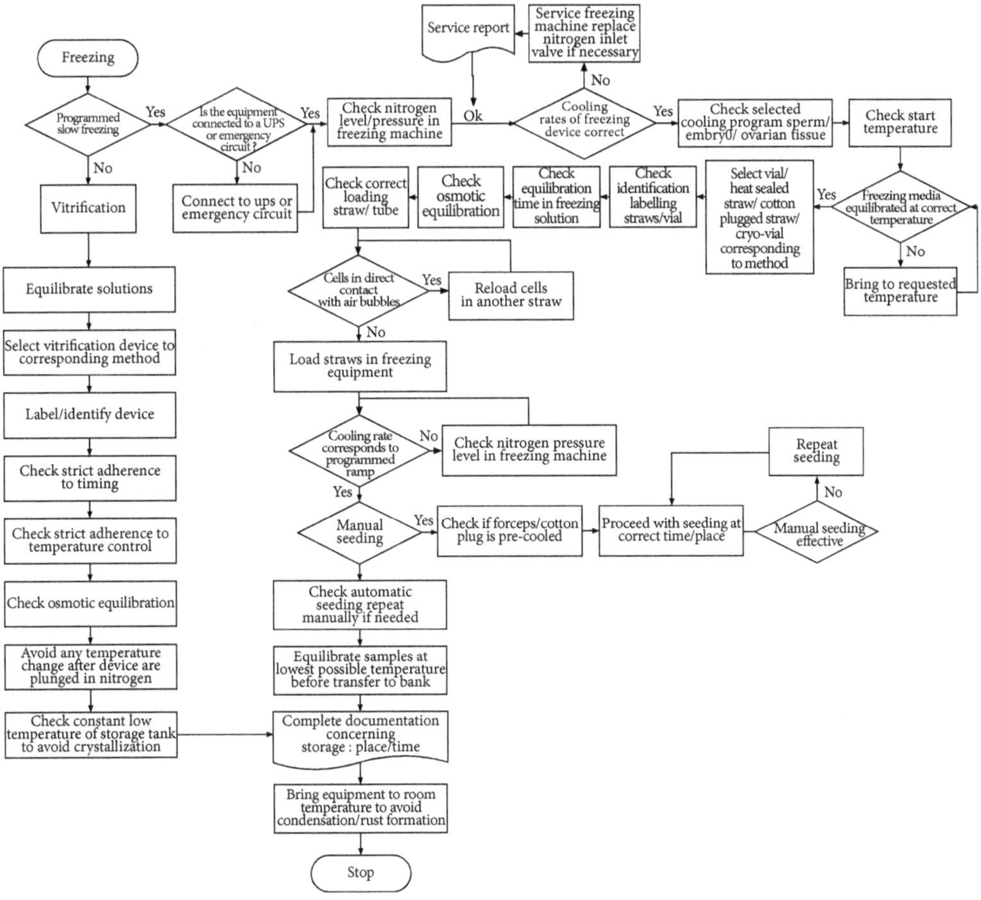

Fig. 10.6 Cryopreservation troubleshooting flowchart.

with frozen material can be helpful in making sure that communications from patients always receive the appropriate attention.

A checklist to cover all eventualities is helpful in conducting a cryostore audit, including the question "Have the patients responded to any communication or sent any new instructions regarding consents?" If any part of the checklist for disposal cannot be completed, further investigation is required before removing any cryopreserved material from the cryostore. Extra care during an audit is required for:

1. Patients who have the same name: all frozen material must have a unique identifiable code, usually with three identifying features (full name, date of birth, unique clinic accession number); appropriate checks/witnessing will decrease the chance that cryomaterial for the wrong patients is ever removed from storage.
2. A change in consent to provided extended storage.
3. A change in consent for removal of gametes/embryos from storage and allowing them to perish.
4. A couple separate, one partner requests continued storage and the other requests termination of storage. Individual gametes may be stored accordingly, but problems

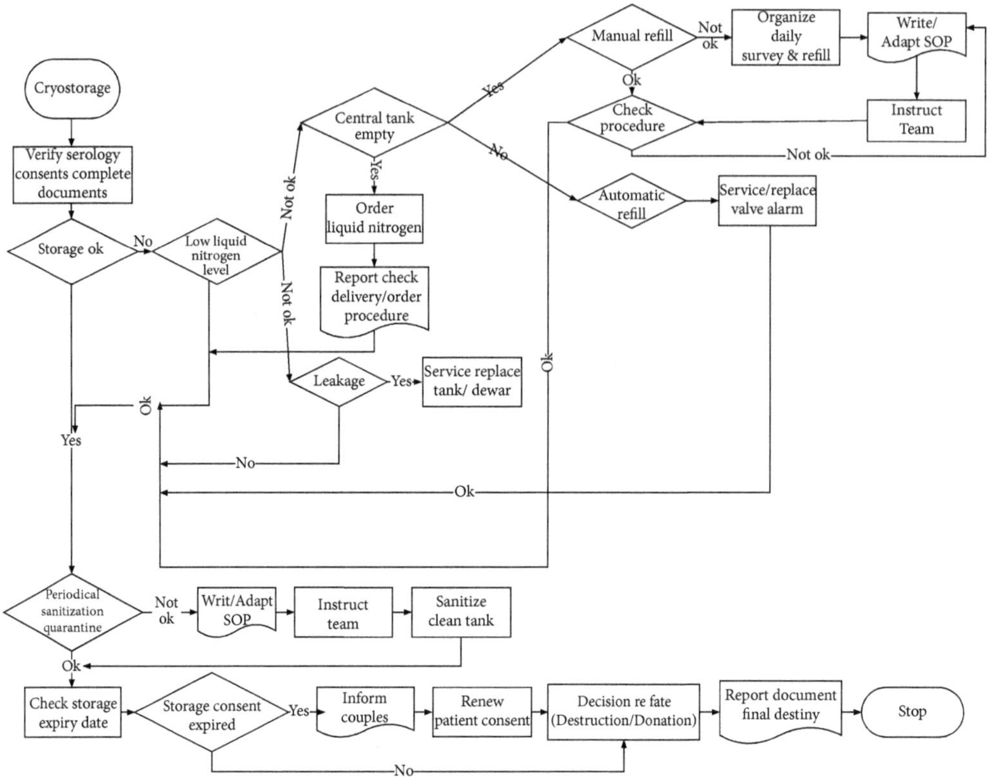

Fig. 10.7 Cryostorage troubleshooting flowchart.

arise in the case of embryos. In the UK, a "cooling off" period of one year is recommended to see if the couple can agree on a united decision; if they cannot agree, storage of cryopreserved embryos must be terminated.

5. Couples approaching the end of their defined storage period should be contacted in advance, so that they have sufficient time to consider the options (usually to remove the gametes/embryos from storage, extend storage, or donate the material to a licensed research project).
6. Couples may have several gametes/embryos in storage generated in different treatment cycles. The gametes/embryos from specific dates must be used in accordance with the patients' wishes. Cryostorage from different dates should be highlighted in the patients' notes.
7. When auditing a donor cryobank, the rules for the donors relating to anonymity and any special requests from the donor should be followed. This includes specifications from the donor regarding duration of the storage period. Note that the issue of payment to donors and the number of families that can be created from a single donor can vary for different countries: any clinic importing donor gametes or embryos should check that payment issues and the number of families produced from that donor are acceptable to the importing country legislation.

Box 10.3 Religious and ethical considerations in the final disposition of embryos

Couples may sometimes request burial for their zygotes/embryos at the end of the storage period, and this may require a special "funeral" SOP, with designated location for the burial. One clinic invited a clergywoman to compose a brief ceremony, and legal advice was obtained regarding how to proceed. Since destruction at the end of the storage period is imposed by law, the cells had to be destroyed before handing over to the couple. Given the associated emotional impact, an SOP for destroying the cells was also generated, so that the couples could be reassured that the cells were completely (and atraumatically) lysed (dissolved) by placing them in sterile distilled water. The SOP should respect all religions of the patients who have cryopreserved embryos.

8. Members of the embryology team change; staff should always remain up to date with all information relating to the contents of the cryostore.
9. An audit must be carried out without distraction – the task of removing embryos from storage requires full concentration.
10. Do not attempt to audit too much in a specific period of time: regular breaks are needed in order to maintain full concentration for an accurate audit.

Mistakes can be avoided by establishing a system that includes robust checks before any gametes/embryos are removed from storage. Straw labels should have a unique cryonumber for the couple, and labels from thawed straws can be attached to the lab report for the thaw.

Further reading

Association of Clinical Embryologists – guidelines on good practice in clinical embryology laboratories 2012. *Hum Fertil (Camb)*, **15**(4), 174–89.

Brison D, Cutting R, Clarke M, Wood M (2012). ACE consensus meeting report: oocyte and embryo cryopreservation. *Hum Fertil*, **15**(2), 69–74.

Byrd W (2002). Cryopreservation, thawing, and transfer of human embryos. *Semin Reprod Med*, **20**(1), 37–43.

Elder K, Dale B (2010). In *In Vitro Fertilization*, 3rd edn, Ch 12. Cambridge University Press, UK.

Letur-Körnisch H, Collin G, Sifer C, *et al.* (2003). Safety of cryopreservation straws for human gametes or embryos: a study with human immunodeficiency virus-1 under cryopreservation conditions. *Hum Reprod*, **18**(1), 140–4.

Tucker MJ, Morton PC, Sweitzer CL, *et al.* (1995). Cryopreservation of human embryos and oocytes. *Curr Op in Obstet Gynecol*, **7**, 188–92.

Tucker M, Liebermann J (2007). *Vitrification in Assisted Reproduction: A User's Manual and Trouble-shooting Guide*. CRC Press.

References

Arav A, Revel A, Nathan Y, *et al.* (2005). Oocyte recovery, embryo development and ovarian function after cryopreservation and transplantation of whole sheep ovary. *Hum Reprod*, **20**(12), 3554–9.

Bernal P, Chang C-C, Nagy PZ (2014). Oocyte and embryo cryopreservation media. In *Culture Media, Solutions, and Systems in Human ART*, ed. Quinn P, Cambridge University Press.

De Santis L, Coticchio G (2011). Theoretical and experimental basis of slow freezing. *RBM Online*, **23**(3), 290–7.

Edgar DH, Gook DA (2012). A critical appraisal of cryopreservation (slow cooling versus vitrification) of human oocytes and embryos. *Hum Reprod Update*, **18**, 536–54.

Fahy GM, Levy DI, Ali SE (1987). Some emerging principles underlying the physical properties, biological actions, and utility of vitrification solutions. *Cryobiology*, **24**, 196–213.

Feng GX, Zhang B, Shu JH *et al.*(2010). Effects of artificial shrinkage of blastocoeles before vitrification on pregnancy outcome. *Zhonghua Fu Chan Ke Za Zhi*, **45**(11), 838–42.

Hsieh YY, Tsai HD, Chang CC, Lo HY (1999). Ultrarapid cryopreservation of human embryos: experience with 1,582 embryos. *Fertil Steril*, **72**, 253–6.

Iwayama H, Hochi S, Yamashita M (2011). *In vitro* and *in vivo* viability of human blastocysts collapsed by laser pulse or osmotic shock prior to vitrification. *J Assist Reprod Genet*, **28**(4), 355–61.

Kojima E, Fukunaga N, Nagai R, Kitasaka H, Ohno H, Asada Y (2012). The vitrification method is significantly better for thawing of slow-freezing embryos. *Fertil Steril*, **98**, S124.

Leibo SP, Pool TB (2011). The principal variables of cryopreservation: Solutions, temperature and rate changes. *Fertil Steril*, **96**(2), 269–76.

Min SH, Lee E, Son HH, Yeon JY, Koo DB (2013). Forced collapse of the blastocoel enhances survival of cryotop vitrified bovine hatching/hatched blastocysts derived from *in vitro* fertilization and somatic cell nuclear transfer. *Cryobiology*, **66**(2), 195–9.

Morris GJ (2006). Cryopreservation of murine embryos, human spermatozoa and embryonic stem cells using a liquid nitrogen-free, controlled rate freezer. *RBM Online*, **13**(3), 421–6.

Mukaida T, Oka C, Goto T, Takahashi K (2006). Artificial shrinkage of blastocoeles using either a micro-needle or a laser pulse prior to the cooling steps of vitrification improves survival rate and pregnancy outcome of vitrified human blastocysts. *Hum Reprod*, **21**(12), 3246–52.

Parmegiani L, Tatone C, Cognigni GE, *et al.* (2014). Rapid warming increases survival of slow-frozen sibling oocytes: a step towards a single warming procedure irrespective of the freezing protocol? *RBM Online*, **28**(5), 614–23.

Son WY, Yoon SH, Yoon HJ, Lee SM, Lim JH (2003). Pregnancy outcome following transfer of human blastocysts vitrified on electron microscopy grids after induced collapse of the blastocoele. *Hum Reprod*, **18**(1), 137–9.

Tabata N, Funayama M, Ikeda T, Azumi J-I, Morita M (1995). On an accident by liquid nitrogen – histological changes of skin in cold. *Forens Sci Internat*, **76**(1), 61–7.

Tomlinson M, Morroll D (2008). Risks associated with cryopreservation: a survey of assisted conception units in the UK and Ireland. *Hum Fertil (Camb)*, **11**(1), 33–42.

Vanderzwalmen P, Bertin G, Debauche CH *et al.* (2002). Births after vitrification at morula and blastocyst stages: effect of artificial reduction of the blastocoelic cavity before vitrification. *Hum Reprod*, **17**(3), 744–51.

Chapter

11 Troubleshooting ICSI procedures

Introduction

Micromanipulation has numerous applications in ART, including ICSI, assisted hatching, oocyte or embryo biopsy, fragment removal, and induction of blastocoel collapse prior to vitrification. Transfer of material (maternal spindle, pronuclei, cytoplasm) between oocytes or zygotes is less common, but also requires advanced micromanipulation skills.

A micromanipulation station is made up of an inverted microscope fitted with modulation contrast optics to facilitate visualization in three dimensions, a mechanical heated stage, pipette holders mounted to the left and right of the stage, syringes connected to the pipette holders, and joysticks to move the pipettes during micromanipulation (Fig. 11.1). Problems during micromanipulation procedures can arise not only from all of the variables already described in the other chapters of this book, but also extrinsically from equipment, consumables, and technique, and intrinsically from the inherent properties of the gametes or embryos themselves.

Location of the micromanipulation station

Micromanipulation requires complete concentration, and the station should be located in a quiet area of the laboratory, in a position with minimal vibration and away from other activities. Meticulous attention to detail is required throughout, and the embryologist must not be disturbed during a micromanipulation procedure.

Air flow should be minimal; a plastic shield can be built around the station to seclude the activity and also reduce air flow. An anti-vibration platform or marble/granite slab placed on top of multiple squash balls can help to reduce vibration; these should be placed on a solid base, away from doors and elevators. Anti-vibration tables with a platform installed on air cushions for the microscope, separate from the supports for injectors and manipulators, have the advantage that the microscope is free-standing.

The ergonomics of the seating and working position are important; seats should be sturdy and comfortable for both men and women (*see Case reports*) and at a suitable height relative to the micromanipulator station.

Equipment and supplies

Before selecting micromanipulation equipment, make sure that it is compatible with your existing microscope. Some systems may require an additional fixing plate for certain microscope makes/models. Correct installation is critical: equipment that is not set up and functioning correctly will cascade problems to the micromanipulation procedures.

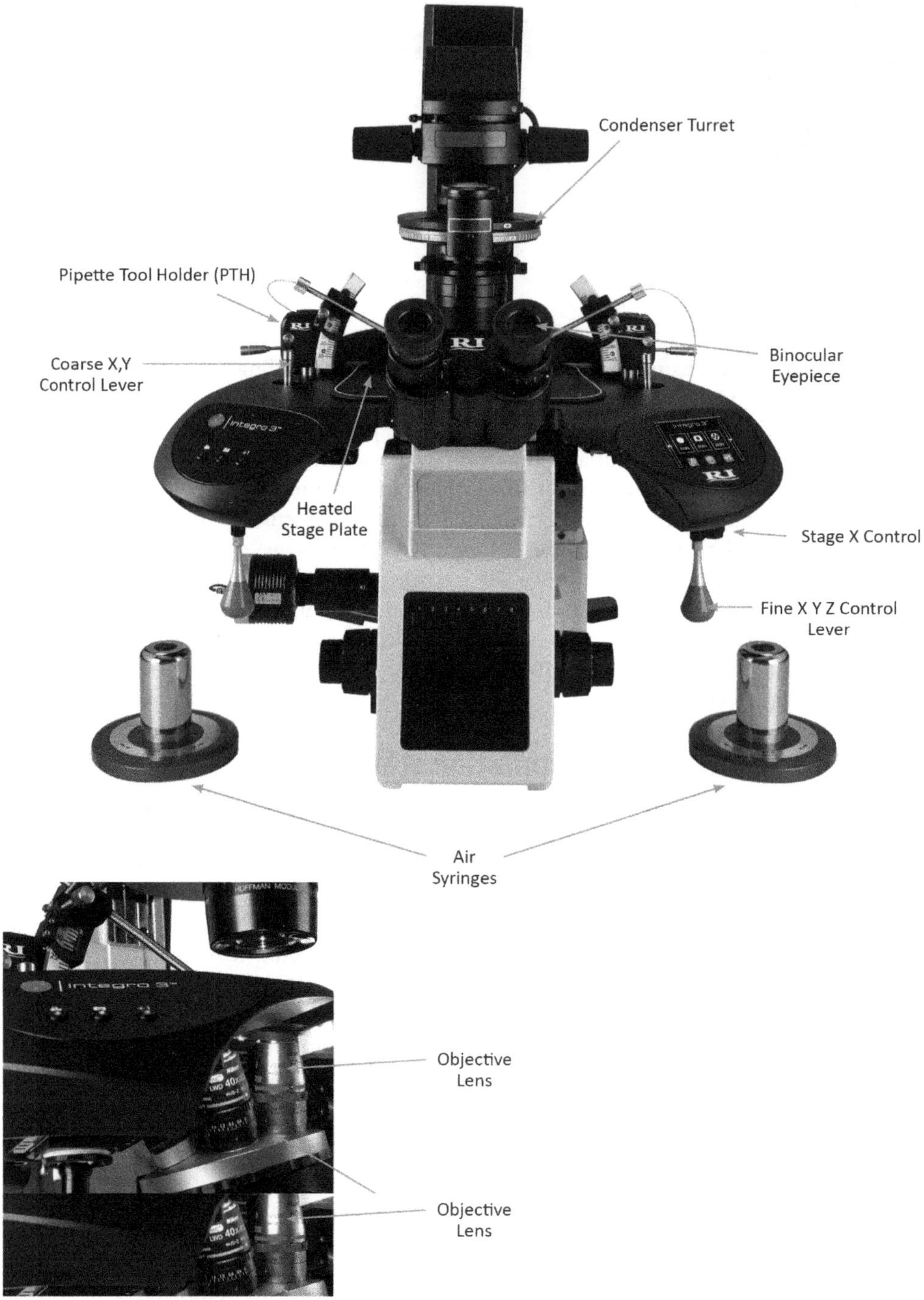

Fig. 11.1 Micromanipulation station (*Integra 3^{TM}*, Research Instruments Ltd), with parts identified. A black and white version of this figure will appear in some formats. For the color version, please refer to the plate section.
(*Image kindly supplied by Research Instruments Ltd, UK. www.research-instruments.com.*)

Box 11.1 Case reports

Ergonomics and vibration can be a source of problems:

1. A busy lab had two adjacent micromanipulation stations, both fitted onto anti-vibration platforms, but sharing the same benchtop. Using just one of the stations was fine, but vibrations increased for both operators if the adjacent station was in use. The solution was to separate the stations onto two separate benches.
2. Another lab had a station positioned on a workbench that was fixed against a solid wall, with an elevator on the other side. The elevator was used only minimally during the day, but its use increased between 4 pm and 6 pm, when staff began to leave the premises. A solution was either to change the location of the station to the other side of the lab, or to perform the micromanipulations before 4 pm.
3. Similarly, a micromanipulation station was fixed to the wall of a building adjacent to a railroad, with a train passing every 9 minutes. The problem was solved by using very efficient "air cushion" anti-vibration tables.
4. A lab that was staffed by female embryologists found that a small "kneeling" chair for ICSI was very comfortable – however, this was less comfortable for a male embryologist joining the staff!

The equipment is usually installed initially by an engineer from the supplying company; it is important to be present at the time of installation in order to learn how the different parts of the system fit together and operate during day-to-day running and maintenance.

Microscope

A basic knowledge of microscopy is essential, including a working knowledge of controls and adaptors for passage of light in all microscopes used in the IVF laboratory. The microscope used for micromanipulation will have additional features such as a modulation contrast optical system. Manuals should be readily available and SOPs written to cover the use of all parts of the equipment (see Chapter 2).

A standard inverted microscope with binocular eyepieces of 10× or 15× magnification is commonly used for micromanipulation. The objective lenses vary from a combination of 4×, 10×, 20×, 40× and 100×, and other magnification objectives may be available on request. It is useful to have the objectives fitted in ascending order on the objective turret, so that a clockwise turn of the thumbwheel will increase the magnification in a step-wise fashion. Ensure that the objectives are completely screwed into the objective turret: an objective that is not fully fitted will not provide optimal magnification. Thoroughly clean the eyepieces and objectives when they are in place – a tiny smudge or piece of dirt (even from a fingerprint) can obscure the optics. It is important to be familiar with the different types of objectives; those to be used for micromanipulation techniques must have a long working distance (LW on the objective), because the glass layer of the heated stage is intercalated between the plastic layer of the micromanipulation dish and the lens. The objectives should be plane-linear and achromatic, so that spherical and chromatic aberration is corrected.

- *Chromatic aberration*: light distortion that occurs when a lens fails to focus all wavelengths of light (colors) to the same convergence point. This can be seen as "fringes" of colour along boundaries between bright and dark parts of the image.

- *Spherical aberration*: an optical effect that occurs when light rays that strike a lens, or are reflected when they strike the edge of a mirror, are refracted to a greater extent than are those that strike closer to the center. This optical effect produces an imperfect image.

Considering the many hours that will be spent using a microscope, spending time and money to ensure a correctly tuned microscope of high optical quality is definitely a worthwhile investment!

An additional 4× objective with a plastic "spacer" tube fitting is available for some micromanipulation rigs. This is a plastic tube of a length that corresponds to the working distance from the highest and lowest point of the fitted pipette, which allows the pipettes to be set up in mid-air above the dish (usually at a height of approximately 14 mm). This has the advantage that pipette tips are less likely to break if set up in mid-air, and the pipette angle can be adjusted so that tips are straight prior to lowering them into the dish. However, always take extra care when lowering the pipettes into the dish for the first time, so that the tips do not touch the base of the dish and break. Remember that the spacer tube is only an approximate guide to the height of the pipette, and may not be the exact distance. Some electric manipulators offer the option of storing positions once the pipettes are correctly installed, including the maximum depth for pipette alignment, so that breakage can be avoided.

Optics

Condensers may be fitted as a turret at the top part of the microscope. Hoffman optics accentuate optical gradients so that the resulting image has different shades of grey. The modulated lighter and darker areas produce a 3-D image, which facilitates micromanipulation. The modulator has a slit design with three zones to control the amount of light transmitted: dark, narrow gray, and clear zones. The condenser turret can be rotated to provide optimal visualization according to the objective in use. This feature is frequently underused, but matching the condenser and objective can improve the micromanipulation procedure.

Hoffman modulation optics have an accessory lens (Bertrand Lens) which can be inserted into the light path above the analyzer. When determining vibration directions or interference figures using parallel or convergent polarized light, this lens is used to bring the image of the interference figures into focus. It is essential to use the appropriate filter for each objective, and each set must be tuned by using the Bertrand filter.

Microscopes with integrated modulation contrast (IMC) are set up so that the orientation of the contrast (direction of the shadow in the 3-D image) is identical for all objectives. This ensures that the spatial impression of the specimen stays the same when changing objectives or magnification. Adjustment of the slit diaphragm is not necessary with IMC.

Nomarski interference contrast optics (also known as differential interference contrast, DIC) uses an illumination technique to enhance contrast in unstained, transparent samples. Light is polarized by a polarizer situated below the condenser, and the polarized light passes through a prism, which causes slight displacement. A second prism and polarizer provide interference contrast, and the resulting shading results in a 3-D image. However, despite the introduction of high-quality plastics, this system can only be used in conjunction with glass-based dishes, and therefore has largely been replaced by the Hoffman system.

If gametes or embryos cannot be clearly visualized in 3-D, the optical modulation settings may need to be adjusted.

Heated stage

Different types of heated stage are available: opaque stages that allow passage of light through them, or metal stages with an aperture for the light source. The stage temperature should be set to maintain microdroplet temperature (and hence the gametes/embryos within them) at 37.0 °C, validated using a thermocouple placed within the microdroplet. The microdroplet temperature will be influenced by the size of the stage aperture (if present), the heat emitted from the light source and the ambient temperature surrounding the equipment (see Chapter 6 for more information on temperature). The heated stage controller should usually be set to a temperature slightly higher than 37.0 °C to compensate for heat loss through the dish base. Heated stage validation should provide an accuracy of +/− 0.1 °C. Some heated stages may be fitted with built-in alarms to provide alert in case of unacceptable temperature variations.

A single temperature setting will suffice only if a single type of dish is used for all micromanipulations. If different types of dishes are used, especially if the material of the dish base differs (e.g. plastic/glass), the heated stage should be reset and revalidated for each dish type (Fig. 11.2). A flow of warmed air below the stage has recently been introduced as an extra feature by some micromanipulation equipment companies, as a means of potentially improving the consistency of heated stage temperature. As with all new innovations, microdroplet temperature must be validated prior to use, to confirm that the air flow does not cause additional evaporation from microdroplets. Since the air flow is below the dish, this should not be an issue as long as sufficient oil overlay is used.

Some microscope stages provide additional peripheral areas for resting dishes in between micromanipulations. Make sure that temperature settings on these areas are validated, and take care that dishes are never at risk of being accidently knocked off the peripheral areas.

An X–Y mechanical stage facilitates precise dish placement and movement between microdroplets. The X–Y controls can differ between stages from different companies: it is important to be completely familiar with the controls, since a clockwise turn of the controller from one company may move the stage left, and the same movement on a different

(a) (b)

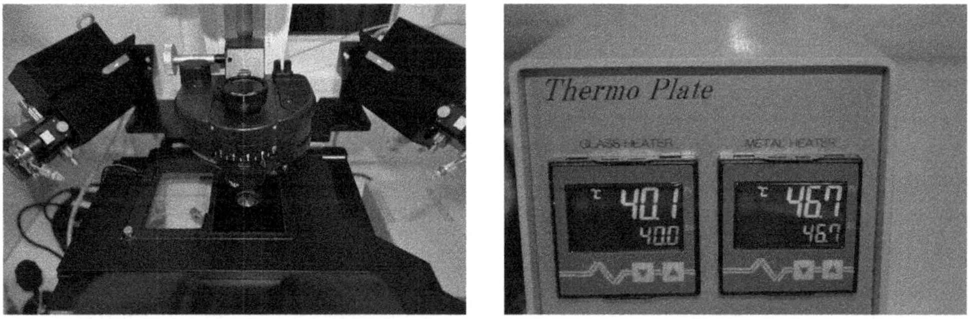

Fig. 11.2 Different types of heated stage may be required for some micromanipulation activities. (a) Micromanipulation system with two heated stages, glass on the left, and metal on the right, with an aperture for the microscope objective. (b) Each stages has a separate temperature controller, which should be validated to ensure that the microdroplet temperature in a dish placed on each stage is maintained at 37 °C. This usually requires different settings for the dual heated stage controller, depending on the ambient air temperature. A black and white version of this figure will appear in some formats. For the color version, please refer to the plate section. *(Images kindly provided by the Hewitt Fertility Centre, Liverpool Women's Hospital, UK.)*

controller may move the stage right! Stages with a mechanical insert for dish placement may have a spring mechanism to secure the dish – dishes must be inserted and removed very carefully. Some microscopes have a hinge that allows the neck and condenser to be inclined backwards, reducing the risk of touching the pipettes when dishes are moved away.

Pipette holders

Pipette tool holders (PTHs) secure the pipettes in the system; taking the PTH apart is a useful exercise, in order to understand how it fits together. The PTH is usually loosened at the tip by a counter-clockwise turn in order to fit a pipette. If oil syringes are used, ensure that a meniscus of oil can be seen exuding from the tip before inserting the micropipette. Pass the micropipette gently through the initial opening, and then through an internal "O" ring. There should be a slight resistance as it passes through the "O" ring; do not use excessive force, as this may cause damage.

Resistance may be caused by a blockage: take the PTH apart and check for remnants from a previous pipette that may have broken inside. Passing a small-gauge needle through the lumen should ensure that the lumen is free. Once the pipette is in position, the tip can be tightened via a clockwise rotation. A pipette may sometimes break during installation, in which case the holders must be cleaned and checked to make sure that no glass particles are left inside; these will impair the gentle control of fluid flow required during manipulations. Oil systems can be flushed with the oil in the micro-injectors; air systems can be cleaned by pushing a fine metal wire through the entire length of the holder until the particles are removed.

The pipette tips must be flush with the base of the dish, and the angle in the pipette is therefore crucial. Commercial manufacturers provide Quality Assurance for the angles of their pipettes, but the angles may need slight adjustment for optimal performance in your system. Fitted angle guides for PTHs can help in setting the tool holder angle, but they should be used as a guide only, as the angle may need further adjustment. Angle guides can also be made in-house to facilitate pipette alignment.

Some PTHs are designed so that the complete holder can be moved laterally/outwards from the area of manipulation; this design facilitates pipette set-up and provides easy access to the dish. When familiarizing yourself with this type of micromanipulation station, be careful not to accidentally catch your hand on any fitted pipettes when removing the dish from the stage! Double tool holders are often used for PGD or fragment removal. These are usually set to the same angle; the manufactured angle of the pipettes may differ slightly, and some can be set independently. The angle of the injection pipette (IP) must be absolutely precise; the tip of the holding pipette (HP) is larger, and can tolerate a wider variation in angle. Spending time to make sure that angles are accurately adjusted will make the micromanipulation procedure easier.

Manipulators

Manipulators can be controlled mechanically, hydraulically (usually via oil) or electronically. Each type provides movement in all directions, operating via a system of joints, pivots, and screws. Familiarity with the entire system is essential, in particular knowledge of how each part of the manipulator links to the next. All systems can provide excellent micromanipulation, but each requires appropriate training. Switching between systems may involve an initial period of adjustment, but this should be relatively simple if the basic skills have been learned.

Fig. 11.3 Eppendorf CellTram Air syringe for micromanipulation, with pneumatic manual microinjector.

Manipulators position the pipette tips precisely, in the X–Y–Z plane. Their joystick controls can be either hanging or upright. Coarse control allows larger movements, and fine control allows the microscopic movements required during the micromanipulation procedure. The level of control in oil-hydraulic systems can be changed by adjusting the movement ratio. All controls should initially be centered to the midpoint of their scales, to allow leeway in all planes. Complete familiarity with the manipulator controls is essential before starting any micromanipulation procedure.

Oil-hydraulic systems have better control, but may be prone to drift, particularly if there is any strain on the tubing. Always make sure that the controls are centered before each new procedure, with a little slack in the tubing.

More recent models include memory buttons that can auto-store pipette position and individual microdroplet position in the dish; these "extras" can potentially reduce the time needed for each micromanipulation. Digital images and procedure timing can also be recorded. Piezo-assisted ICSI has been reported (Takeuchi *et al.*, 2001), but its use to date is more applicable for animal models.

Syringes

Pneumatic (air) (Fig. 11.3) and hydraulic (oil) syringes are commonly used for micromanipulation in ART, and each system has its own advantages/disadvantages. Oil systems must be completely free of bubbles, from the syringe and throughout the tubing. The syringes are easily filled and topped-up via a two-way screw. Electronic microinjectors may be used to inject fixed volumes in pronuclear micromanipulation for transgenics (Fig. 11.4).

Micropipettes

Micropipettes vary in angle and lumen size, and IPs may be spiked or not spiked; the type of pipette used is a personal choice. Keeping a variety of IPs of different lumens in stock for all micromanipulation procedures is beneficial (see Case report, Box 11.2). All micropipettes are very fragile, and great care is needed in removing them from their packaging – make sure that you are familiar with the safest way to do this. Check each pipette to ensure it is not broken: breakages most frequently occur at the tip. Companies do their best to ensure safe packaging and delivery, but there may be a batch with frequent breakage. This must be reported to the company so that the source of the damage can be investigated. Never use a broken pipette, even if the damage is minor.

Laser

Only Class I laser products should be used for clinical micromanipulation, as defined by international laser safety standards (International Electrotechnical Commission – IEC 60825-1). The precision and size of pulses must be carefully selected, to prevent thermal

(a)

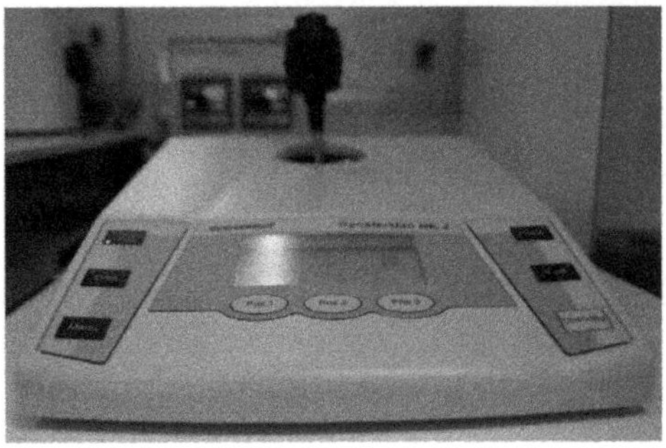

(b)

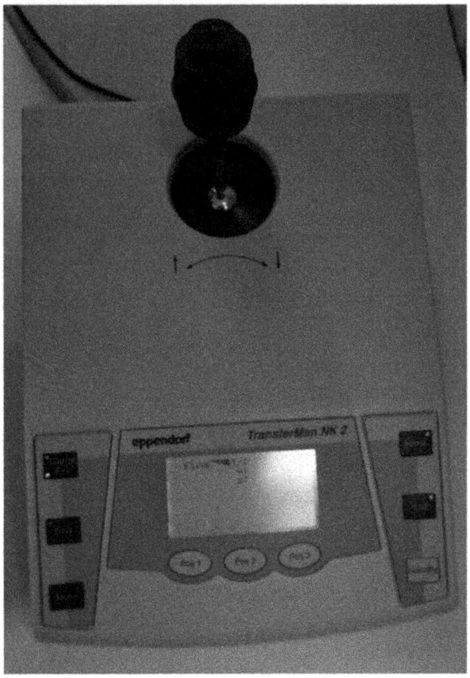

Fig. 11.4 Two views of an electronic micromanipulator system (Eppendorf TransferMan®). This system provides X–Y–Z coarse and fine control, and can also store and recall pipette positions.

Box 11.2 Case report

A couple with severe teratozoospermia were scheduled for ICSI. Following sperm preparation, motile sperm were plentiful, but all of the sperm heads were too wide to be aspirated into the injection pipette. Different aliquots from the prepared sperm all contained sperm with the same morphology. Fortunately, the lab had an alternative type of injection pipette with a slightly wider lumen, and sperm could be aspirated after fitting this new pipette (each sperm head could just fit into the lumen). However, when it came to the injection procedure, the pipette lumen was significantly larger than was routinely used, and thus carried an increased risk of damaging the oocyte during injection. Thankfully, all of the oocytes were injected satisfactorily, and the damage rate seen the next day was less than 5%. Interestingly, the fertilized oocytes had two pronuclei of different sizes, with one pronucleus at least twice the size of the other. This was thought to be due to the size of paternal chromatin from each large sperm head.

damage to any area of the gamete or embryo where the laser is not directed. A laser fitted to a microscope objective can be used to ablate the zona pellucida in assisted hatching or biopsy, to induce blastocoel collapse prior to cryopreservation (Mukaida *et al.*, 2006), for sperm immobilization (Xu *et al.*, 2007), and to assist fertilization (Woodward & Campbell, 2014). Some laser systems allow the position of pulses to be pre-set along a chosen path that the laser will then ablate automatically.

The laser must be tested to confirm that the pulse is directed exactly at the target location, at correct intensity. Some systems may be prone to drift, and need to be recalibrated before each use: using a black marker pen, draw a cross on the upper surface of the base of a micromanipulation dish. Direct the laser to a specific area of the ink cross and fire the pulse. A "white" ablated area should appear on the cross where the pulse has fired. If this is not exactly on the target position and of the desired size, recalibrate the laser. Proceed to clinical use only when target position precisely matches the actual position.

Focusing the microscope

Using a black marker pen, mark an "X" on the upper surface of the base of a micromanipulation dish, as per laser calibration described above. Place the dish on the stage and focus on the "X" with the lowest objective (usually 4×). When this is in focus, change to a higher magnification objective and establish that the "X" can be brought into focus; at higher magnification, this is the edge of the "X." If the pipettes are to be set up in mid-air using a 4× spacer objective, turn the objective turret to this position and adjust each pipette in its PTH until the pipette tip is straight and in focus. Since the dish is already correctly in focus with the 4× objective in the position where the gametes or embryos will be, do not adjust the focus at this stage – only adjust the height and angle of the pipette in the PTH. If there is no spacer objective, set up the pipettes above the dish without using the microscope. With experience, the angles of the pipette tips can be set up in approximately the correct position. The pipettes can be set up initially using a dish with medium alone, without oil, and then rinsed and tested for correct function/control prior to the injection procedure.

Lower the HP until it is just above the dish. The tip should be slightly out of focus, but the HP tip can be aligned so that it is parallel to the surface of the dish. Do the same for the IP, noting that this is a more fragile pipette. Slowly lower both pipettes so that they are in focus, and make final adjustments to their positions. Final adjustment can also be accomplished by focusing the microscope on the interface between oil and medium in the sperm droplet, either with direct contact between oil and medium, or with a small amount of air between oil and medium. Finalise pipette alignment by lowering them into a microdroplet of medium, and equilibrate by aspirating medium into the pipettes. The volume to be aspirated is a matter of personal preference, but the medium meniscus should not go beyond the position where the bevelled angle changes. It is important to equilibrate the pipettes in a microdroplet under oil before starting any micromanipulation.

Extensive practice in pipette installation and sperm micromanipulation (tail breaking and catching) is essential before any ICSI procedures are attempted; a spherical algae colony of approximately the same size as a human oocyte (Volvox) can be used for "oocyte" training (Just *et al.*, 2004). Microscopic beads can be used to mimic the injection procedure. Unfertilized or immature oocytes can be used for ICSI training only if there is written consent from both partners, stipulating that these oocytes will never be used for treatment.

For biopsy training, unfertilized or immature oocytes can be used for polar body biopsy, and either embryos derived from abnormal fertilization or supernumerary embryos that are not considered suitable for treatment can be used for blastomere biopsy. As with all training that involves gametes or embryos, written patient consent (from both the sperm and oocyte provider) is required, stipulating the type of training and confirming that this will not affect a couple's treatment in any way.

Media

ICSI media

Buffered media (e.g. HEPES or MOPS) is commonly used for micromanipulations in order to maintain stable pH outside of the incubator. Manipulations can also be performed in the same media that is used for gamete/embryo culture, as long as the duration of the micromanipulation procedure is relatively short, in order to avoid adverse changes in pH. It is essential that the medium is free of any precipitate that might be aspirated into the injection pipette and may be injected into the oocyte. The number of oocytes/embryos held in a dish for micromanipulation depends upon the skill of the embryologist: it is recommended that no more than six oocytes/embryos are held in each micromanipulation dish. Procedures should take no more than 10 minutes; if more time is required, fewer oocytes/embryos should be transferred to the micromanipulation dish. Optimal oocyte numbers with respect to timing can be assessed by measuring Key Performance Indicators such as fertilization and implantation rates.

Polyvinyl pyrrolidine (PVP) and hyaluronan

PVP is commonly used as a viscous medium for sperm manipulation prior to injection. Commercial preparations vary from 7% to 10%, and can either be used directly, or further diluted with buffered medium. Injection dishes with PVP and medium droplets under an oil overlay can be prepared at the beginning of the working day, but preparation closer to the ICSI procedure time avoids potential changes in viscosity. PVP is a viscous solution, which aspirates into a pipette slowly, with equally slow release into the dish – take care to make sure that the full required volume is released into the dish. Adverse effects of PVP on sperm membranes and zygote cytoplasm have been reported (Kato & Nagao, 2012; Parmegiani *et al.*, 2014), and although sperm capture in medium alone is possible, this requires enhanced micromanipulation skills to control sperm in less viscous solutions. Sperm should be exposed to PVP for a minimal period of time, and the injection procedure should carry the smallest possible volume of PVP into the oocyte.

Hyaluronan (hyaluronic acid, HA) is a natural and readily degradable glycosaminoglycan that can be used as a substitute for the artificial PVP polymer without jeopardizing the outcome of the treatment cycle (Barak *et al.*, 2001; Balaban *et al.*, 2003; Worrilow *et al.*, 2013). Hyaluronan can be metabolized by the oocyte (personal communication Yves Ménézo), and may have an additional role in sperm selection. Recombinant hyaluronan is available as SpermSlow™ (Origio, Denmark).

Hyalurodinase

Hyaluronidase solution (Hyase™, Vitrolife, Sweden) is available at a concentration of 80 IU/ml, or as a10× concentrate for dilution with buffered medium on the day of use. This solution is used to remove the cumulus cells surrounding the oocyte (denudation). Prepare a dish containing droplets of Hyase in buffered media under an oil overlay, and warm the dish to 37 °C. The hyase droplet volume should be large enough to allow transfer of cumulus–oocyte complexes (COC) using a Pasteur pipette; residual cumulus cells will remain in the droplets after oocytes are removed. Volumes of 100–200 µl are adequate for five COCs; prepare the appropriate number of dishes according to the number of oocytes for denudation. Manipulate the COCs in each dish swiftly and precisely, observing all oocytes in the droplet throughout the denudation procedure. A Pasteur pipette can be used to tease off the majority of the cumulus cells. Oocytes with residual attached cumulus cells are then transferred to a buffered medium droplet for further denudation, either with a single denudation pipette or with a series of denudation pipettes with decreasing lumen diameter (290 μm, 170 μm, 155 μm, 135 μm). Avoid damaging the oocyte while removing cumulus cells, and wash the denuded oocytes through a series of medium droplets to remove residual Hyase solution.

Denudation can also be quickly accomplished in a four-well dish with Hyase solution in one of the wells. Up to ten oocytes kept in another well can be transferred in and out of the Hyase with a Pasteur pipette for 30 to 45 seconds, and then transferred to a fresh dish with buffered medium for further mechanical cumulus removal.

The amount of cumulus to remove also needs consideration: since incomplete denudation may enhance embryo quality (Ebner *et al.*, 2006), removing only sufficient cumulus cells to visualize the polar body and safely perform ICSI may be beneficial. However, when time-lapse technology is used, adequate imaging necessitates removal of the majority, if not all, of the cumulus cells.

Timing of cumulus removal after oocyte collection varies. Oocytes may benefit from remaining in contact with cumulus cells to allow full cytoplasmic maturation, even if the metaphase II stage has been reached. Patrat *et al.* (2012) investigated the optimal timing for oocyte denudation and ICSI, and concluded that 2 hours is an optimal time between oocyte retrieval and denudation, and that ICSI should take place as soon as possible after denudation. The timing should be established and validated in your own laboratory, and may need to take into account factors such as:

1. The oocyte number: extra time for the ICSI procedure will be required if there are a large number of metaphase II oocytes to inject.
2. Oocyte maturity and sperm preparation: if the sperm is cryopreserved, it is a waste of resources to prepare the sperm in advance of the denudation, in case no oocytes are suitable for ICSI.

Dish type and microdroplet arrangement

Dishes with an inner non-toxic and "embryo-friendly" coating are available. In Europe, there has been a move away from using dishes that are not CE-marked, and Falcon 351006 dishes are no longer the dish of choice. Falcon 353655 and Nunc 150265 dishes are a similar size and are suitable for use. Standard 60 mm dishes (Falcon 353652 or Nunc 150270) can be used, but larger dishes with higher sides will require steeper angles for the IP and HP (40° or more).

Using the lid of a 60 mm diameter dish as an inexpensive alternative is not recommended. Cell culture plastics are treated in order to change their surface tension and promote better cell adhesion; lids are generally not cell culture treated, and a lack of surface tension will result in flatter droplets.

Microdroplet arrangement for ICSI is a matter of personal choice. Each dish usually contains a sperm microdroplet (e.g. PVP or HA) and several media microdroplets for the oocytes. An extra microdroplet of medium can be used for pipette equilibration. In order to maintain consistent temperature, all droplets should be placed away from the periphery, closer to the center of the dish. Pipettes are less likely to hit the sides of the dish in this position. Numbering the media microdroplets on the underside of the dish helps in recording

Box 11.3 Case report

"On my first day of working at a clinic, whilst preparing to denude oocytes using Hyase, the local embryologist suggested that we go for a break after moving the cumulus–oocyte complexes into Hyase. I explained that the denudation process had to be completed once started. Apparently, a previous locum embryologist always transferred the cumulus–oocyte complexes into Hyase and left them in the acid whilst he took a 5-minute break. On his return, the oocytes were completely denuded, and he washed them through media and returned them to the culture dish. However, the ICSI results were understandably very poor. Exposing oocytes to Hyase for 5 minutes has a detrimental effect on the oocytes, and the exposure should be minimal.

This highlights the fact that the competence of locum embryologists should be assessed, in order to avoid introducing poor practice."

Box 11.4 Case report

A clinic was concerned about low ICSI success rates. Investigation revealed that more than ten oocytes were transferred to the ICSI dish, and the embryologist spent at least 5 minutes carefully positioning each oocyte with the polar body exactly at the 6 o'clock position. The entire duration of the procedure, using a single dish, was over 1 hour. After following a recommendation to reduce the number of oocytes to two per dish and spend less time on exact polar body positioning, the success rates increased. Eventually, they increased the number of oocytes to six per dish, and were able to complete each ICSI in less than 10 minutes.

which oocytes have been injected. Micromanipulation dishes containing HEPES-buffered media can be equilibrated in a CO_2 incubator if the lids can be tightly sealed on to the dishes. Alternatively, they can be equilibrated in a non-CO_2 incubator set to 37 °C (see Chapter 5).

Sperm samples with large amounts of cellular debris, e.g. from samples with very low counts or after surgical retrieval, need a separate droplet arrangement in order to assess the whole sample and locate motile sperm. A pipette can be used to create line droplets of media for the sample, or to "draw together" separate round media and sperm droplets. Motile sperm often swim to the droplet edge. In difficult cases, it is useful to first locate the sperm in a separate dish, leaving the oocytes in the incubator. In order to avoid potential damage to sperm membranes, pool the located sperm together into a separate media droplet instead of PVP, without immobilizing them. Individual sperm can then be moved to the PVP droplet and immobilized immediately prior to microinjection.

Troubleshooting tips

Dish and pipette preparation

1. If microdroplets of different solutions are to be placed in the same dish (e.g. Hyase and medium), place the droplets at a distance sufficient to prevent mixing. However, it is useful to position the droplets so that they can all be viewed under a 4× objective.
2. Remove all bubbles when preparing the dish and after transferring gametes or embryos to microdroplets.
3. Do not incubate dishes containing buffered media in CO_2 incubators unless the lid can be tightly sealed to the dish.

4. Never overload a dish with a higher number of oocytes than can be safely dealt with. Remember to minimize the time oocytes are exposed to manipulations during denudation and micromanipulations. During denudation, restrict the number of COCs to a safe number (a maximum of five at a time is recommended).
5. When fitting a pipette, never force it into the holder. Check to see if the lumen is blocked by remnants of a previous pipette that may have broken.
6. If the IP has bubbles on the outside, bring the IP tip next to the HP tip. Carefully inserting the IP tip into the lumen of the HP tip should remove all bubbles.
7. When adding sperm to the PVP droplet, only add a very small volume (1 μl) at the edge of the droplet. If a large volume is added, the PVP will be diluted and lose its viscosity, and the sperm motility will not be reduced. If a high concentration of sperm is added, catching individual sperm will be difficult.
8. If the motility of sperm in the final preparation is very poor, add the sperm to a media droplet for selection, and then transfer to the PVP for tail breakage.
9. If sperm cannot be trapped with the IP, the IP tip might not be parallel with the dish base. Slightly readjust the angle of the IP, so that the "toe" of the IP point is lower towards the base.
10. Placing the tip of the pipette just on the interface between medium and oil, with the rest of the pipette in the oil, will stop sperm, and they may even swim into the IP by themselves. This is a very useful trick for testicular tissue preparations with a large amount of debris that can block the pipette.

Micromanipulation

1. Carefully think through the ergonomics of the micromanipulation procedure. It is worth investing in an adjustable chair with good lumbar support.
2. Syringes should be positioned relative to the microscope so that their handling is comfortable; this is a matter of personal choice. The "motorbike stance" has the HP on the left tool holder, with the corresponding syringe to the right of the microscope; the IP is in the right tool holder, with the corresponding syringe to the left of the microscope. Alternatively, both syringes can be placed on the same side.
3. Make sure that all of the controls are in the central position before starting a micromanipulation procedure. Coarse and fine controllers should be central, and syringes set at halfway between their limits. If a range of movement limit is reached during manipulation, return gametes/embryos to optimal culture conditions, re-centralize equipment, and start again.
4. Working at the limits of the range means that the equipment may be more prone to changes in tension, which will have an impact on micromanipulation.
5. Keep all equipment clean: dust settling into joints can prevent further movement and create problems. Clean very carefully, so that disks or screws are never accidentally maladjusted. Motorized manipulators may need repair if this happens. Some motorized manipulators may suffer from shaky movements; this is caused by inadequate electric wiring, and earth connections must be checked.
6. Keep all of the equipment lubricated: lack of lubrication can cause joints to wear.
7. When multiple users carry out procedures at the same station, always check the equipment prior to setting up for a micromanipulation. Knowing how to adjust all settings is critical, especially if the previous user has changed them.

8. Taut tubing between connecting syringes and/or electrical and hydraulic tubing can affect control. Always allow a little slack in tubing, and never put the equipment under strain.
9. Oil inside tubing often remains in place for long periods, and can be fully changed periodically. Oil is usually clear; any change in color to yellow or brown is a sign of deterioration. Different types of oil can be used: silicone oil has the highest viscosity, mineral oil is less viscous, and Fluorinert™ has the lowest viscosity. Air bubbles in the oil will interfere with syringe pressure, and should always be removed.

ICSI procedure

1. Cut the sperm tail only once; multiple "aggressive" cuts can lower the quality of resulting embryos after ICSI (Velaers *et al.*, 2012). The single cut should make a "V" shape in the tail.
2. Sperm that have not been adequately "cut" can retain motility. Although motility should stop relatively soon after injection, do not inject a motile sperm, as the tail will disrupt the cytoplasm.
3. If the selected sperm cannot be aspirated into the IP, choose another sperm; don't spend too much time trying to aspirate a single sperm.
4. Do not proceed to ICSI with a sperm head that is detached from the tail, as injection of the tail is essential: unlike the mouse system, the human oocyte needs a sperm centrosome for spindle formation. Select another sperm for ICSI.
5. Changes in fluid density may cause the sperm to move up the IP during transfer from PVP to medium droplet. As soon as the pipette tip is in the medium droplet, apply slight positive pressure to the IP syringe to return the sperm to the tip before moving to the oocyte.
6. ICSI can be successful performed by injecting the sperm head-first or tail-first (Woodward *et al.*, 2008), but head-first injection allows the sperm to be precisely positioned at the very tip of the IP immediately prior to oolemma insertion.
7. Multiple sperm can be aspirated into the IP, spaced equally apart, but this carries a risk that sudden aspiration of ooplasm into the IP might encompass two sperm cells. One sperm at a time is the safest method.
8. Do not spend too long trying to align the polar body at either the 6 o'clock or 12 o'clock position. The polar body is only an approximate marker of spindle position, and its positioning is not that critical (Woodward *et al.*, 2008).
9. Insert the IP into the oocyte exactly at the equator. If both oolemma and IP tip are in crisp focus, they should both be at the same height above the base of the dish. Injecting an oocyte in a hemisphere greatly increases the chance of oocyte damage. The position of the IP prior to injection can be tested by gently pushing against the oolemma and moving the IP up and down with fine control. Equal amounts of the oocyte should be observed on either side of the IP.
10. If the oolemma does not rupture, even with significant aspiration, stop negative pressure, gently release the oolemma back to its original position and withdraw the pipette. If the sperm is still in the pipette, attempt the oolemma rupture again: the "stretched" oolemma should be more prone to rupture on the second attempt.

Box 11.5 Case report

Inter-practitioner checks revealed that one individual showed a significantly higher oocyte damage rate after ICSI when compared with other colleagues. The senior embryologists had a one-to-one discussion with the individual, asking for a description of the entire ICSI process. This was followed by observing the embryologist during an ICSI procedure, with digital camera recording via the microscope. Troubleshooting the complete process led to the discovery that this embryologist carried out one step differently: instead of immediately returning cytoplasm to the oocyte along with the sperm after aspiration and membrane rupture, they continued to aspirate cytoplasm further up the IP for another couple of seconds. This additional action caused further stress to the oocytes, and a significant proportion did not recover. The embryologist was retrained, so that cytoplasm was replaced immediately after membrane rupture. The next 20 ICSIs were monitored, and demonstrated a much lower damage rate. At the following year's inter-practitioner checks, the damage rate was consistently low for all of the team.

11. Immediately apply positive pressure to return the ooplasm (and sperm) as soon as the oolemma ruptures. Do not aspirate further ooplasm into the IP, as this will disrupt the ultrastructure of the oocyte and increase the risk of damage.
12. If the membrane on the opposite side of the oocyte is aspirated during cytoplasmic aspiration, stop! Gently return the contents of the oocyte, withdraw the pipette, and return the oocyte to the droplet of medium. Repeat the ICSI with this oocyte after all the remaining oocytes have been injected. Position the oocyte with polar body at a different position to the original one, so that a different area of the oolemma can be entered. Rupturing the opposite membrane during ICSI carries a high risk of lysis.
13. If the sperm "sticks" to the IP tip as the IP is withdrawn from the oocyte post-ICSI, and is clearly not within the ooplasm, release the oocyte from the HP. Perform ICSI on all other oocytes in the dish, and then return to the original oocyte to re-attempt ICSI. Try to aspirate an alternative area of the oolemma, by rotating the oocyte so that the polar body lies at a different position.
14. Some patients are prone to producing oocytes that are particularly fragile, and these have an increased lysis rate after ICSI. A clinical trial showed that using a laser to ablate a channel in the zona prior to insertion of the IP did not improve lysis or fertilization rates (Richter *et al.*, 2006). The use of lasers for pre-selected patients with fragile zonae has not yet been fully evaluated; reducing compression on the oocyte by zona drilling may have a role.
15. Occasionally, a clear membrane rupture is observed and confirmed by aspirating cytoplasm into the IP, but the sperm has been injected into a "fold" in the cytoplasmic membrane and not through ruptured membrane into the cytoplasm. This is equivalent to subzonal injection with an immobilized sperm, and results in low fertilization; the sperm can sometimes be found under the zona. For correct injection, the sperm should be seen to "dive" into the oocyte.

Patient management

1. In cases that require surgically retrieved sperm, the surgical retrieval can be performed in advance of the ovarian stimulation cycle, with cryopreservation of retrieved sperm.

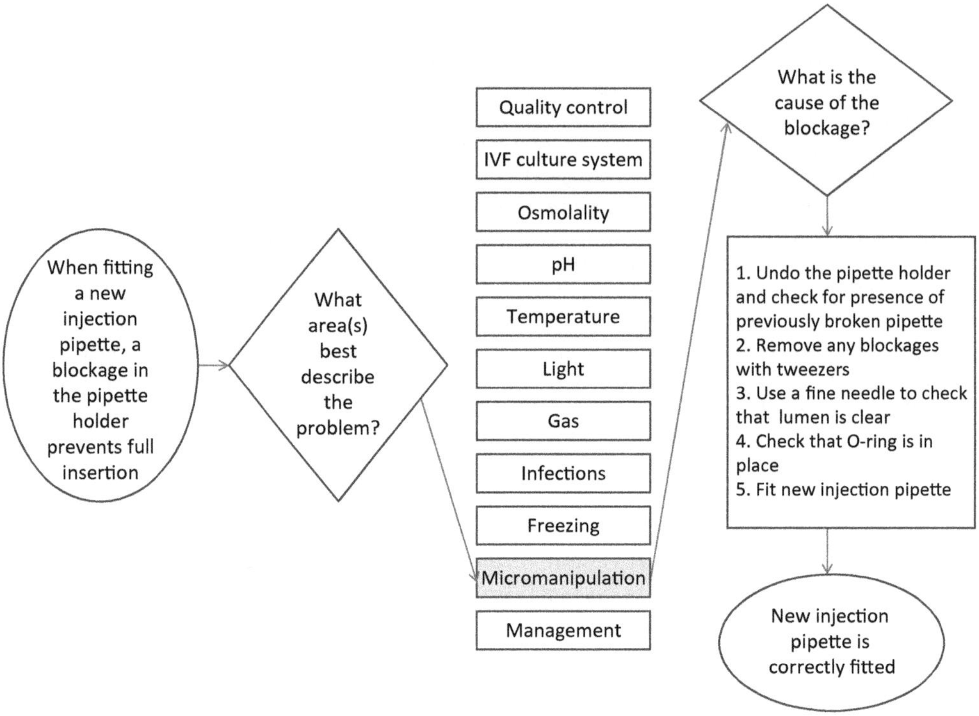

Fig. 11.5 Process map for troubleshooting tool holder problem.

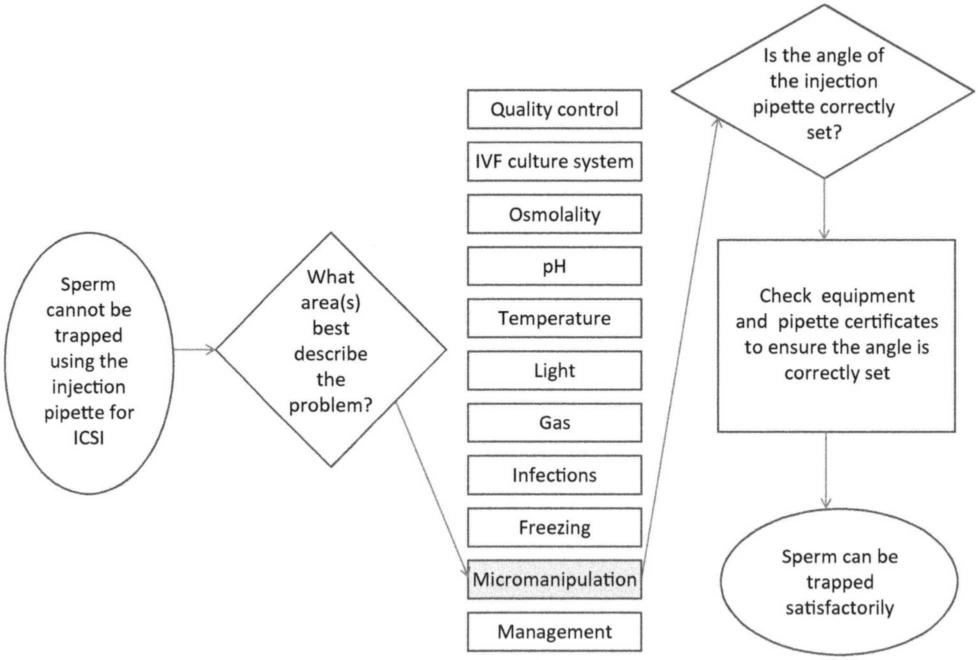

Fig. 11.6 Process map for troubleshooting problem with sperm capture.

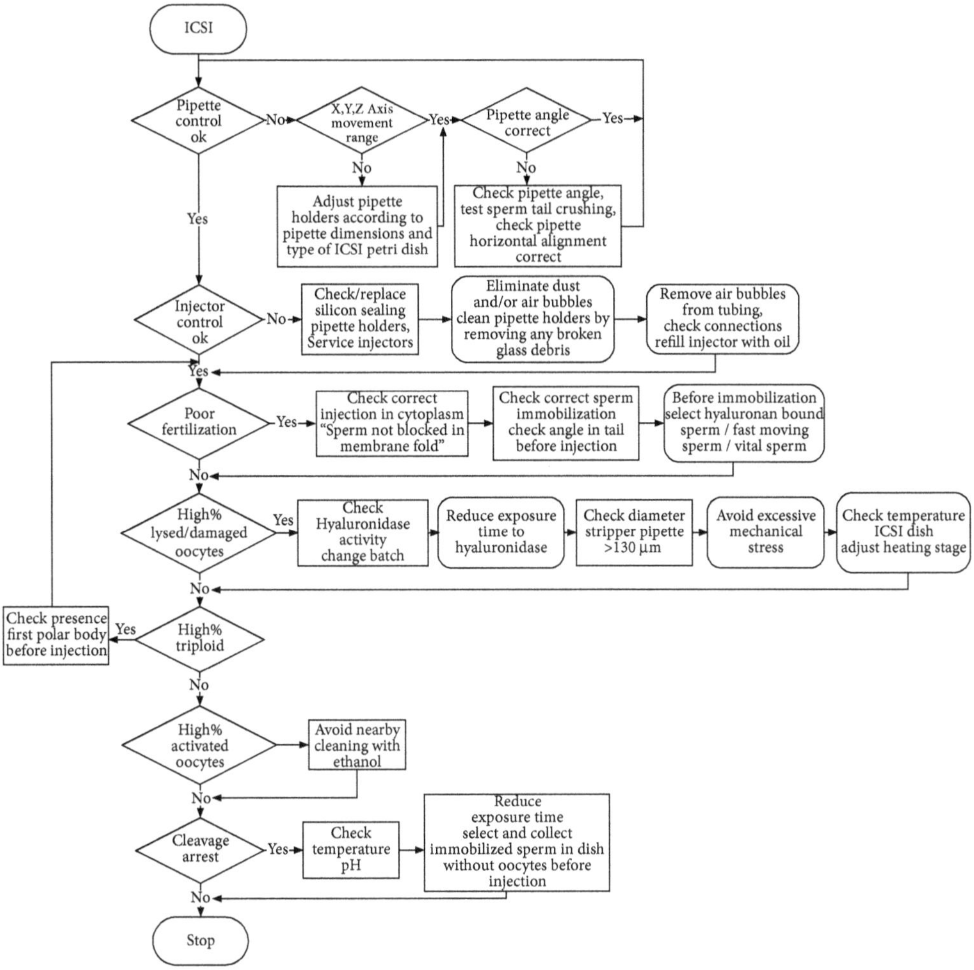

Fig. 11.7 Flowchart for micromanipulation troubleshooting.

The couple then have assurance that sperm are available for ICSI, with the additional benefit that both partners do not have operations (and anesthesia) on the same day. Allocation of matched donor sperm, with signed consent for its use, is a useful back-up in case sperm are not found after thawing the specimen.

2. PESA should be performed bilaterally before considering more invasive options such as TESA/TESE, unless there is clinical reason for performing unilateral sperm retrieval.
3. In cases that are scheduled for IVF, it is beneficial to carry out semen assessment and sperm preparation prior to the time of oocyte retrieval, in case ICSI is indicated instead of IVF. If conversion from IVF to ICSI is necessary due to poor quality sperm "on the day," both partners must provide written consent for ICSI; the female partner cannot legally sign a consent if anesthesia has been administered.

References

Balaban B, Lundin K, Morrell JM, Tjellström H, Urman B, Holmes PV (2003). An alternative to PVP for slowing sperm prior to ICSI. *Hum Reprod*, **18**(9), 1887–9.

Barak Y, Ménézo Y, Veiga A, Elder K (2001). A physiological replacement for polyvinylpyrrolidone (PVP) in assisted reproductive technology. *Hum Fertil*, **4**(2), 99–103.

Ebner T, Moser M, Sommergruber M, Shebl O, Tews G (2006). Incomplete denudation of oocytes prior to ICSI enhances embryo quality and blastocyst development. *Hum Reprod*, **21**(11), 2972–7.

International Electrotechnical Commission. IEC 60825-1: 2007 (2nd edn 2007 I-SH 01).

Just A, Gruber I, Wober M, Lahodny J, Obruca A, Strohmer H (2004). Novel method for the cryopreservation of testicular sperm and ejaculated spermatozoa from patients with severe oligospermia: a pilot study. *Fertil Steril*, **82**(2), 445–7.

Kato Y, Nagao Y (2012). Effect of polyvinylpyrrolidone on sperm function and early embryonic development following intracytoplasmic sperm injection in human assisted reproduction. *Reprod Med Biol*, **11**(4), 165–76.

Mukaida T, Oka C, Goto T, Takahashi K (2006). Artificial shrinkage of blastocoeles using either a micro-needle or a laser pulse prior to the cooling steps of vitrification improves survival rate and pregnancy outcome of vitrified human blastocysts. *Hum Reprod*, **21**(12), 3246–52.

Parmegiani L, Cognigni GE, Filicori M (2014). Sperm selection: effect on sperm DNA quality. *Adv Exp Med Biol*, **791**, 151–72.

Patrat C, Kaffel A, Delaroche L, *et al.* (2006). No advantage of laser-assisted over conventional intracytoplasmic sperm injection: a randomized controlled trial [NCT00114725]. *J Exp Clin Assist Reprod*, **3**, 5.

Richter KS, Davis A, Carter J, Greenhouse SJ, Mottla GL, Tucker MJ (2006). No advantage of laser-assisted over conventional intracytoplasmic sperm injection: a randomized controlled trial. *J Exp Clin Assist Reprod*, **3**, 5.

Takeuchi S, Minoura H, Shibahara T, Shen X, Futamura N, Toyoda N (2001). Comparison of piezo-assisted micromanipulation with conventional micromanipulation for intracytoplasmic sperm injection into human oocytes. *Gynecol Obstet Invest*, **52**(3), 158–62.

Velaers A, Paternot G, Debrock S, D'Hooghe T, Spiessens C (2012). Triple touch sperm immobilization vs. single touch sperm immobilization in ICSI – a randomized trial. *Reprod Biol Endocrinol*, **10**, 65.

Wolf JP, Fauque P (2012). Optimal timing for oocyte denudation and intracytoplasmic sperm injection. *Obstet Gynecol Int*, Article ID 403531.

Worrilow KC, Eid S, Woodhouse D, *et al.* (2013). Use of hyaluronan in the selection of sperm for intracytoplasmic sperm injection (ICSI): significant improvement in clinical outcomes – multicenter, double-blinded and randomized controlled trial. *Hum Reprod*, **28**(2), 306–14.

Woodward BJ, Campbell KHS (2014). Laser assisted IVF: a simplified technique to improve zygote production rates without the need for ICSI. *IVFLite*, **1**, 12–16.

Woodward BJ, Montgomery SJ, Hartshorne GM, Campbell KH, Kennedy R (2008). Spindle position assessment prior to ICSI does not benefit fertilization or early embryo quality. *RBM Online*, **16**(2), 232–8.

Woodward BJ, Campbell KHS, Ramsewak S (2008). A comparison of headfirst and tailfirst microinjection of sperm at intracytoplasmic sperm injection. *Fertil Steril*, **89**(3), 711–14.

Xu ZP, Sun HX, Hu YL, Zhang NY, Zhao X (2007). Laser-assisted immobilization causes no direct damage to sperm DNA. *Zhonghua Nan Ke Xue*, **13**(3), 216–18.

Chapter

12 Managing risk in an IVF laboratory

Introduction

Chapter 1 introduced this book with a reminder of Murphy's Law:

> *"If there are two or more ways of doing something, and one of them can lead to catastrophe, then someone will do it."*

Everyday events of any nature always have some element of associated risk; the complex and sophisticated systems involved in successful ART inevitably involve a huge catalog of potential risk factors, and a cascade into catastrophe can only be prevented/circumvented by acknowledging the risks, cataloging the elements surrounding them, and establishing a system of active risk management that recognizes and acknowledges influence and responsibility for events in the future.

An ART setting is subject to risk from a variety of sources: quite apart from "Acts of God"/external disaster, risk factors can be contributed by clinical, laboratory, legal/regulatory, administrative, and even moral/ethical situations. Building and managing an effective laboratory team is crucial in minimizing risk in the laboratory: lack of knowledge, training, and supervision all contribute to risk, as well as other factors such as physical ineptitude, stress, inadequate or inappropriate equipment, or conditions. Management decisions, organizational processes or "corporate culture" may lead to conditions that might be prone to error or procedure violation. Whether unintentional or deliberate, human beings will on occasion make errors, break rules or violate protocols in a manner that requires intervention. Human failures are more likely to take place in an environment where inadequate organizational policies can lead to inappropriate decisions.

A comprehensive Quality Management System (QMS) is a first-line defence against potential risk, encompassing training, protocols and procedures, good systems of communication and feedback, witnessing, and back-up systems. Active risk management can identify failures in organization/systems by keeping meticulous records that can be objectively examined on a regular and routine basis. Introducing new techniques or changing existing protocols always involves risks; these can be minimized by making sure that it is not a unilateral decision, and that everyone is aware of and has agreed to the procedure.

The Human Fertilization and Embryology Authority (HFEA), the legislative/governing body for clinics that provide IVF treatment in the UK, has developed a "Risk Tool" that acknowledges the importance of effective QMS. All licensed IVF clinics regularly submit comprehensive reports to the HFEA, including incident reports, changes in personnel, as well as full details of every cycle, from start of stimulation to final outcome. The data submitted is used to assess all clinics in terms of their compliance with statutory requirements, as well as the quality of their service. Compliance with all of the requirements of the HFEA Code of Practice and Directions is the key to preventing situations of risk (http://www.hfea.gov.uk/6674.html):

> *"The Risk Tool acknowledges that actually "doing things" to prevent risks happening is critical. While effective documentation is important, having a good standard operating procedure (SOP) is only effective in helping prevent risks when the SOP is followed. Therefore, an audit of practice ensuring that SOPs are being followed contributes more to the mitigation of risk than simply having an SOP or a quality indicator."*

The HFEA Risk Tool identifies six key areas:

- Incorrect identification of gametes or embryos
- Consent failures
- Damage or loss of gametes or embryos
- Cross-infection of gametes, embryos, or patients
- Multiple pregnancy
- Incorrect or incomplete information on donors.

All of the principles outlined in Chapter 2 of this book are important in minimizing the incidence of risk situations in the IVF laboratory; effective team management is one of the most important contributing features.

Managing an IVF team

ART is a field of human endeavor that uniquely and completely depends upon multidisciplinary cooperation and collaboration: clinical medicine, embryology, genetics, scientific research, nursing, counseling, ethics... and politics. Something worthwhile and beautiful, the birth of healthy babies, can be achieved only by concerted focus on a common goal, replacing individual ambition with coordinated teamwork and collaboration: all members of the team working together to reach a common purpose.

The delicate and sophisticated process of nurturing a new life "*in vitro*" is a highly stressful situation, not only for the tiny preimplantation embryo, but also for the embryology team who hold the responsibility of supporting and maintaining its viability before transfer to a natural maternal environment. As outlined in each chapter of this book, numerous critical events and parameters are involved in successful *in vitro* culture, all of which must be coordinated with precise attention to detail. Successful transfer of a viable embryo ultimately depends upon a chain of events that requires very close teamwork, with trust and interdependence between all members of the team working together. Equivalent to the links of a chain – a single weak link can lead to danger/collapse, jeopardizing the treatment cycle.

An optimal IVF laboratory requires careful management to ensure that workload is safe, staff training with regular review is effectively in place, work is allocated according to established and recognized competencies, and that any interpersonal issues that arise are handled sensitively and effectively.

Safe workload

The number of cases that can be safely handled by an embryology team in order to achieve optimal results and effective patient treatment must be based upon the range and complexity of treatments offered, and the levels of training and experience of available staff. An ART laboratory that handles ancillary andrology and/or diagnostic services, embryo biopsy, and cryopreservation requires a different staffing level from one that has access to any of these as a separate facility. A caseload that exceeds the capacity of the facilities and staff inevitably puts gametes and embryos at risk, subjecting them to fluctuations in temperature/pH due to increased opening of incubator doors, particularly since incubators have a limited number of spaces available.

"What is the optimal case number per embryologist per year?" is a question commonly raised, in particular by managers who need to handle budgets and decide on staffing levels. Figures of 100 to 150 cases per embryologist are often quoted, and there are moves towards an international consensus. However, it is unfair to imply that "one embryologist can cover 100 or 150 cycles/year" as too many variables are involved, including individual competence and how the workload is organized. A clinic that batches and performs 100 cycles during 1 month in a year carries out approximately the same number of cycles as one that performs two cycles/week, but the situations are completely different. One embryologist would struggle to perform the former, but should be able to handle the latter more easily. Different strategies can be used to make effective use of staff and resources in a manner that will optimize positive outcomes. For example, cycles can be scheduled so that a group of IVF/ICSI cases is handled during a 2-week period, and this is followed by 2 weeks of treatment that require less laboratory involvement (e.g. IUI, FER), allowing the laboratory to catch up on paperwork, administrative, and maintenance tasks. Similarly, fresh cycles can be scheduled over the period of a month, followed by a month of FER/IUI cycles only. These strategies can improve working conditions, reduce occupational stress and the risk of errors, ultimately minimizing the chance of substandard patient care.

Management of regular supplies and equipment maintenance is crucial; a system that delegates clearly defined maintenance, ordering, routine preparation, and cleaning tasks to specific and appropriate personnel will help to ensure that all supplies are available as required, and everything runs as smoothly as possible without risk of equipment or supply failure. Laboratory work can be separated into areas that can be managed by staff of a lower grade, i.e. "technicians" instead of "embryologists", as is the current norm in some countries. Basic routines such as stock control, dish preparation, cleaning, paperwork, etc. can be allocated to technicians. Levels of training vary, and some "technicians" may be competent to perform oocyte collections and sperm preparation; some clinics even have ICSI technicians. However, a trained clinical embryologist/scientist must always be available to supervise such staffing arrangements, make decisions, liaise with the multidisciplinary team and patients – and troubleshoot. As outlined in Chapter 2 (and confirmed by the HFEA), regular audits are essential, incorporating equipment and system checks as well as monitoring of Key Performance Indicators.

Daily fluctuations in the number of cases, especially on weekend days when fewer staff may be available, can be a considerable source of stress. Treatment cycles that involve maximal complexity and time (e.g. those that include surgical sperm retrieval or embryo biopsy/aneuploidy screening or preimplantation genetic diagnosis (PGD) should always be carefully managed with programmed scheduling to make sure that appropriately trained personnel are available, without compromising requirements for other cases on that day. Excessive workload fluctuations can lead to a downturn in results: embryologists become tired, and there may be a temptation to take "short cuts" in order to handle all the tasks that must take place throughout a busy day. This temptation must always be avoided, and SOPs followed precisely without fail; they have been composed and validated for a specific reason. The other members of the fertility team must also be taken into consideration with respect to workload: every case involves clinicians, nurses, anesthetists, and administrative staff, each requiring adequate staff and resources in order to make their contribution safely and effectively.

All technicians and embryologists must have demonstrated competence in all of the tasks that they are scheduled to perform, with frequent assessment and appraisal. A minimum number of cases should be completed by each practitioner to ensure that skills are maintained; this is an essential component of QM according to ISO certification and accreditation norms. Senior embryologists who move to more managerial positions often have less opportunity for lab-based work; if they perform fewer than the proposed minimum number of cases (e.g. participate in less than 25% of the total activity for a particular technique or manipulation), they should either demonstrate competence before undertaking a procedure, or remove themselves from the rota.

Staff training

Training is often performed in-house according to rules specific to the individual clinic. Even the most academically qualified individual will have to undergo practical training in order to perform competently in an IVF laboratory; a skilled, experienced embryologist who changes clinics will have to adapt and re-train to ensure that the new clinic's SOPs are followed.

Some countries have national training schemes, as well as post-graduate qualifications. A training logbook is usually kept by the trainee to provide a record of the tasks completed, with an area to note any mistakes, together with the corrective action taken. Mistakes provide an excellent learning opportunity, as long as corrective action ensures that the mistake is avoided in the future: "Anyone who has never made a mistake has never tried anything new." (Albert Einstein).

After sufficient training, the trainee should demonstrate that he/she can competently perform a task, under full supervision by a senior member of staff. A minimum of 10 to 20 consecutive repeats for a task is usually required before competence is accepted. A signed training agreement between trainee and supervisor will help to ensure that both understand what must be undertaken and accomplished. Patient consent is required in some countries (e.g. UK) in order to allow trainee embryologists to practise using gametes and embryos. Trainees should always respect the nature of the gametes and embryos, and act accordingly, throughout training and in all clinical practice.

As skill level increases, so does the risk of making a mistake. Delicate transfers of embryos require more skill than moving cumulus–oocyte complexes during egg collections. Nervousness may be translated into shaking hands, and this may be exaggerated when dealing with embryos. Holding breath whilst transferring embryos between dishes is a common feature that should be avoided; breathing calmly throughout all embryology procedures is to be recommended. The trainee should always have the "safety net" of an experienced embryologist present to supervise. An embryo may sometimes be "lost" during pipetting or transfer between dishes, and this requires a calm approach in searching for it. The embryo can usually be found in the system somewhere, either in the dish or pipette. This can be a demoralizing experience for a trainee, but the supervisor should emphasize that to err is human, it can happen to anyone. Good supervision during training should ensure that there is minimal risk to gametes and embryos. Supervisors need to be aware that trainees differ, both in technical skills and in personality, and different styles of communication may be required. Authority as a supervisor is best obtained by providing help and assistance, rather than by taking over from the trainee.

Progress during training must be documented, with separate records for all learned techniques and manipulations that include the date, case number, and supervisor's signature at the end of each session. The records should provide verification that the number of procedures and results are within pre-defined limits before the trainee starts any independent work. This type of data collection can be continued after the training period for use during internal competency review audits.

Competency assessments should be recorded for trained embryologists as well as for trainees, with at least annual review (it may be useful to invite an external embryologist to help with a lead embryologist's competency assessment). Review of all competency records can be used for internal quality assurance: if an individual's record shows a lower KPI for a particular task, then a troubleshooting exercise is indicated to assess the individual fully when performing the task. An atmosphere of openness and "no-blame" is important, so that any re-training is seen as a positive action. Competence can also be evaluated with the help of online Quality Assessment Schedules such as UKNEQAS (www.ukneqas.org.uk) and FertAid (www.fertaid.com). This has the additional benefit of allowing comparisons with other centers. A period of re-training is also necessary when returning to work after any prolonged absence (e.g. maternity leave). The skill set should return relatively quickly: five cases may suffice.

An accredited clinic will also have a system of appraisal and competency assessment in place for all members of the clinical and nursing teams; this is an essential component for troubleshooting OCR or ET procedures. The Quality Manager is responsible for assessing KPIs, and all members of the team should be willing to cooperate in re-training as required, with the ultimate goal of improving success rates.

Continuing professional development

A fully trained embryologist should keep up to date with latest developments in the field of clinical embryology, from their own regulatory body as well as from national and international societies. Recent developments, peer-reviewed manuscripts, or conference reports can be presented and discussed at regular team meetings. Interdisciplinary team meetings that include clinicians, nurses, paramedics, counsellors, and embryologists can also be used to cascade changes to guidelines or information from regulatory bodies.

Continuing professional development (CPD) schemes that allocate points for CPD activities are available at national and international levels. In the UK, the Royal College of Pathologists (RCPath) and the Association of Clinical Embryologists (ACE) host their own CPD schemes. The European Society for Human Reproduction and Embryology (ESHRE) hosts a Continuous Embryology Education Credit scheme (CEEC, http://www.eshre.eu/Accreditation-and-Certification/Certification-for-embryologists/Continuous-Embryology-Education-Credits.aspx), and the American Board of Bioanalysts (ABB) in the USA enables their members to collect continuous education units by participating in peer reviewed meetings, including webinars (http://www.aab.org/aab/More_Educational_Programs.asp).

Box 12.1 Case report

A supervisor had set a rule for oocyte collection training, stipulating that 20 oocyte collections had to be performed efficiently and without missing an oocyte. Each dish of each collection was re-checked by the supervisor throughout this period. The rule had an additional caveat: if the trainee missed an oocyte, he/she had to go back to the beginning, and perform a further 20 collections. The training went well, and no oocytes were missed – until the nineteenth oocyte collection. The supervisor found an extra oocyte at the end of the case, to the amazement/shock of both supervisor and trainee. However, the rules were followed, and the trainee performed a further 20 collections without missing an oocyte. Both parties were then certain that the trainee was now able to competently carry out oocyte collections.

Box 12.2 Case report

Evaluation of QA reports for semen analysis revealed that one of the technicians systematically overestimated the number of motile sperm, and in particular those with highly progressive forward motility (Class A, World Health Organization, 2010; Tomlinson *et al.*, 2012). The supervisor observed the individual during the next QA exercise, and noted that the technician was counting the majority of the rapidly moving sperm twice. The situation was corrected after additional training.

On-call

A member of the embryology team needs to be on-call at any hour of the day, in case of emergencies. Despite careful management to ensure reliable equipment performance, there are occasions when equipment fails: an incubator may run out of gas or fluctuate in temperature (either too high, or too low), or a storage dewar may leak nitrogen. All critical equipment should ideally be linked to a 24-hour monitoring service that can either alert the hospital reception or transfer to an auto-dialler that alerts a member of the embryology team out of hours. Only staff who are trained in use of all of the critical equipment may be included in on-call rotas, and specific SOPs should cover all eventualities that detail actions to be taken during an on-call emergency, including any necessary corrective action that can be carried out within proper limits. If external monitoring that can be accessed from home is in place, checking the source of the alarm online can help to decide whether it is a critical event that requires a visit to the laboratory (see Chapter 2, Figs. 2.2–2.5). On-call staff must be able to travel to the IVF laboratory within a defined period of time, ideally as soon as possible; a limit of 20 minutes is acceptable. Adding staff who live at a greater distance to the on-call rota is not practical. On-call staff need to abide by the same rules regarding drugs and alcohol that apply during working hours, and must not be "under the influence" at any time during the on-call period. Additional pay is usually provided to compensate for the inconvenience.

As a safety precaution, on-call staff should always be accompanied by a member of the security team when accessing the IVF clinic during out of hours. If a nitrogen storage vessel has leaked, oxygen levels must be checked before the cryostore is entered (see Chapter 10). If embryos or frozen material need to be moved between incubators or dewars, legal requirements in some countries (e.g. UK) mandate that a second member of the laboratory team is available for witnessing.

Box 12.3 Visitors to the IVF Lab

IVF laboratories frequently host visitors (medical/university students, nurses, doctors, etc.). Visitors should be asked to sign a confidentiality agreement, as patient identities/notes may be visible to them. It is essential that all visitors abide by rules when in the laboratory:

- No wearing of perfume/cologne, body powder, or scented body lotions (see Chapter 8, VOCs).
- Remove make-up and watches or bracelets.
- Disciplined conduct, without distracting any working embryologists or technicians at any time.
- Equipment must not be touched.

When observing any surgical or embryo transfer procedure, they should be asked to alert someone immediately if they feel at all faint, and try to step away (if possible) from any work area where gametes or embryos are being handled.

Team dynamics

Good teamwork results from appropriate selection of good team members and leaders, and therefore maintaining a good team requires careful consideration. A mature and flexible approach, with personal commitment to a joint vision of success for the clinic and the patients is the foundation of a good team. All roles within the team should be clearly defined, and members encouraged to communicate their opinions for ways to improve the service, with support throughout and incentives for rewarding good performance.

Poor teams often have a bad mix of team members, leading to unfortunate dynamics. Lack of guidelines, without proper evaluation of success or competence, and language problems can all have a negative impact on the team. Any situation that results in conflict among staff or psychological bullying should be referred to the Human Resources/ Personnel department for advice when necessary. When there is a change in team leader, e.g. if a good team leader takes maternity leave or a new post, their replacement must be chosen with care, either by promotion from within the current team or by external recruitment. In either case, performance and effectiveness need to be monitored, so that the team continues to work together smoothly. Good performance by a team deserves recognition, both for the efforts of the team as a whole, and for individual members. Ultimately, all members of the team are crucial to success, and each must be valued (Table 12.1).

Communication

The interdependent nature of IVF teamwork imposes a need for mutual trust and respect between all members of staff, at all levels. Situations that raise concerns regarding reliability/trust, "fairness", or competence affect the entire team, and can be a cause of considerable stress and strain that may ultimately jeopardize clinical results. Every individual differs with respect to physical, mental, and emotional resources: someone with very poor three-dimensional perspective may find that the coordination required to carry out an ICSI case quickly and effectively imposes a stressful/ unbearable emotional burden. Staff training and appraisal should ensure that each member of the team is not only competent and capable of carrying out his/her allocated tasks, but also has the emotional stamina and stability to handle the workload. Laboratory managers do not necessarily have the

Table 12.1 The eight essentials of dynamic teamwork (*adapted from Choi & Pak, 2007*)

T	Team
E	Enthusiasm
A	Accessibility
M	Motivation
W	Workplace
O	Objectives
R	Role
K	Kinship

Box 12.4 Case report

An embryology team lost their senior embryologist, who was replaced by promoting a junior member of staff to the role of senior embryologist; this person was the only member of the team keen to take on the extra responsibility. The team had already been forced to adjust to the loss of an experienced team member, and dealing with the same number of cases became difficult. The newly appointed senior embryologist decided that the best way to deal with the workload was to cut staff lunch breaks. This led to ill-feeling within the team, and also contravened the contracts of employment. Feelings within the team eventually came to a head, and an inter-disciplinary team meeting was convened. It was decided to reduce the workload until a full team complement was available. Lunch hours were restored and the team dynamics sorted themselves out. The new senior embryologist had initially made a poor decision, but he has since gone on to be an excellent leader of a full team.

background or training to handle interpersonal conflicts that may arise: classical problem scenarios have been described by Karpman (www.karpmandramatriangle.com), and these models can provide a logical initial approach. A classical "Drama Triangle" can be described among three different types of personality, who take on the roles of Victim–Rescuer–Persecutor.

1. A "victim" may feel that he/she is helpless and blameless ("I can't deal with this by myself"), and not responsible for the problem ("I'm sorry, there's nothing I can do"). They expect change to happen, the problem to be solved by someone else – but are then unable to adapt to any changes that occur.
2. The "rescuer" (very rewarding role); this type of personality assumes that the victim is not able to solve his/her own problem, and offers help without being asked. They often "help" in situations where others could carry out the task themselves.
3. The "persecutor" seldom considers him/herself to be part of the problem, and "accuses" others. This type of personality often manipulates the situation in order to gain from it themselves.

This type of Drama Triangle can be seen in a different perspective, with realization and acknowledgment that:

- The victim is not really as helpless as he/she feels
- The rescuer is not really helping
- The persecutor does not have a valid complaint.

Satisfactory conflict resolution should ensure that each "player" gets their unspoken/unconscious psychological needs met in a manner that they feel justified; there is no need to acknowledge any dysfunction or harm that has been done. Unless the situation is dealt with promptly and effectively, there is a danger that the roles may escalate to:

1. Dependency: the victim has their needs met by letting the rescuer take care of them; the victim receives attention, but no real help. This can evolve to a danger that the victim can switch to being a persecutor, but will perceive this as being accidental or a result of their being misunderstood.
2. The rescuer (a less obvious role) does not actually resolve problems, as there is a motive behind his/her helpfulness. This carries the danger that the rescuer can switch to being a persecutor: "You never appreciate my help."

The triangle has quantitative variables:

- Flexibility – a shift from words to actions, to body language
- Tenacity – some people are extremely persistent
- Intensity – players may be relaxed/passive – or very aggressive.

The type of conflict may have three different categories:

- Socially acceptable
- Undesirable but without irreparable damage
- Drastic harm.

Resolution involves finding a way to deprive the actors of their pay-off; the triangle is converted to a "Winners' Triangle," facilitating changes so that:

1. The "victim" starts to actively seek help, with guidance towards:
    - Solving their own personal problems
    - Concentrating on competences, not on inadequacies
    - Seeking evidence of fair treatment
    - Awareness of the fact that everyone is equally vulnerable
    - Remembering their own successes and achievements.
2. The "persecutor" becomes a negotiator:
    - Using energy for changes, not for control
    - Being persistent, not merciless
    - Fair trade is better than war
    - Formulating goals and wishes, and talking about them.
3. The "rescuer" becomes a helper:
    - Using the competences of others to let them solve their problems
    - Learning to say NO
    - Not taking over without being asked
    - Not doing what others can do
    - Concentrating on the resources of others.

In any situation of conflict, it is important to initiate mediation and start addressing the issues as soon as possible: the longer the conflict continues, the harder it is to resolve. Friedrich Glasl has proposed a model to describe conflict escalation with time (Glasl, 1999); if the problem is tackled sufficiently early, there is greater opportunity for resolution, either internally, or with the help of external professional moderation. As time goes on, loss of face and strategies of threat may start to arise, and if these are not addressed promptly, the situation will escalate to a level that may require external professional mediation before the players plunge "together into the abyss" of irreconcilable differences.

Crisis management

Some parts of the world are more prone to the perils of natural disaster than others: "Acts of God," which are beyond the scope of human preventive action. Severe weather, earthquakes, tsunamis, volcanoes, forest fires, and civil disturbance can all pose a threat to the safety of gametes and embryos in the laboratory, as well as to staff and patients. IVF clinics in perilous geographical locations should make their own assessment of risk levels, and have a policy of "crisis management" in place, with detailed contingency plans to protect patients and staff, as well as the laboratory environment and stability of equipment as much as is possible in case natural (or man-made) disaster should strike.

Contingencies for prolonged loss of utilities (power, water, gas, telephone, IT), loss of staff in case of epidemic illness or travel/transport difficulties should be considered. An on-site generator should ideally be available for back-up power to the laboratory, and as with all critical equipment, tested regularly to make sure that it works, with ongoing monitoring of fuel levels/supplies. Maintaining an uninterruptable power supply (UPS) unit, ready to use in case of disasters, is recommended. UPS units vary in the amount of power they can store: some are used continuously as a back up to counter any fluctuations in power, or during changes from the main grid to the back-up generator. Other UPS units provide more power, but only on a finite basis, and their use should therefore be gauged to ensure back-up for critical equipment only. UPS units tend to be very heavy, and should be handled with care; always bear in mind that their limited power should be kept in reserve for critical equipment.

Electrical connections

Electrical connections must be handled with care: overloaded plug sockets and adaptors linked to further adaptors create unnecessary danger that could be avoided with careful risk assessment. Plug sockets should have only one plug fitted and the network for all critical equipment should be linked to the generator power supply. Sockets linked to a back-up generator may be color-coded to distinguish the "safe" sockets. A benchtop incubator linked in a power loop that is fed from one socket creates the risk that an electrical fault

Box 12.5 Case report

Whilst setting up a clinic in the Caribbean, the power from the national grid became an issue. When international cricket matches were shown on the television in the evenings, the end of the match coverage was always accompanied by a change in power! This fluctuation in electricity supply created a hazard for the critical equipment; UPS units had to be purchased and put into constant use, to ensure a continuous supply of power.

Box 12.6 Case report

ISO certification required that all of a laboratory's incubators should be placed under continuous surveillance. The hospital did not agree to direct phone connections to the on-call embryologist, but instead installed a connection to the emergency desk telephone. This receptionist was then obliged to call the hospital on-call technician, instead of the on-call embryologist. The on-call hospital technician could be either in the hospital, or could be present within 30 minutes. The laboratory was certified with this system in place, but the laboratory directors were not happy with this situation. A decision was made to reduce the risk for patients by adding extra incubators so that each incubator held no more than two cases, and all incubators were placed on an emergency power circuit.

Box 12.7 Case report

In 2011, the Great Eastern Japan Earthquake and resulting tsunami hit many clinics in northern Japan. It has been reported that 12 clinics lost at least one embryo. Clinics can protect their equipment from minor tremors (see Fig. 12.1) but larger tremors will cause inevitable damage. Interestingly, for the Japanese clinics, all the frozen embryos stored in liquid nitrogen dewars remained safe (Ishihara & Yoshimura, 2011).

with one incubator will cascade to the remaining incubators in the loop. The solution is one socket per incubator (Box 12.6).

Troubleshooting during power loss

1. Make sure that the generator is fully serviced and is connected. Check the estimated amount of power that can be provided.
2. Have UPS units ready to connect to critical equipment in case the generator fails.
3. Bring all cultured embryos from different patients together into one incubator, so that only one incubator needs to receive a power source. Check that all dishes are appropriately labeled, and group any multiple dishes belonging to the same patient together.
4. Contact local IVF clinics to enquire if they are unaffected by the power loss, and whether they are able to accept any gametes or embryos transported to their facilities. Gametes and embryos should be transported in tubes only (never dishes), using transport incubators. Clearly label all tubes and check that gametes and embryos are fully traceable to the respective patients.
5. If an eventual power loss is inevitable, vitrify all oocytes and embryos for future use.
6. Document all action and alert any regulatory bodies and patients.

In areas that are prone to earthquakes, cryopreserved gametes and embryos can be stored temporarily, and the temporary stores transported at scheduled intervals to a permanent store in a location that is less prone to seismic activity (Box 12.7, Box 12.8).

If "safety" is theoretically interpreted as "without risk," then this is a situation that does not exist in the daily life of an IVF embryologist. The ISO 14971 norm presents a more realistic definition of "safety": "freedom from *unacceptable* risk." Risk evaluation or rating

Box 12.8 Case report

In 2012, the Practice Committees of the American Society for Reproductive Medicine and the Society for Assisted Reproductive Technology (2012) published recommendations for developing an emergency plan for all IVF programs. These included the need to have contingency plans in place to protect fresh and cryopreserved gametes, embryos, and tissue and to provide for continuation of patient care in case of emergencies or natural disasters. A few months after the publication, Hurricane Sandy hit New York City on October 29 2012. With power loss from both the main grid and back-up generators, one IVF clinic had 90 embryos in their incubators with limited remaining power fed from UPS units. The decision was made to vitrify all embryos, regardless of the stages of development. This safely took place, before the power ran out from the UPS units. Over the following 6-month period, six patients had frozen embryo transfers and all six achieved pregnancies, endorsing the contingency plan with confirmation that safety was not compromised (Goldman *et al.*, 2013).

As a second contingency during Hurricane Sandy, where possible, patient care was continued for those scheduled to attend for IVF treatment at clinics not affected by the tragedy. Emergency cooperation with other fertility clinics outside the affected area enabled patient treatment to continue without cancellation. Some patients chose to continue with their oocyte collections but have the oocytes cryopreserved. In the months following the disaster, the cryopreserved oocytes were subsequently transported to the original clinic for oocyte warming and treatment (Goldman *et al.*, 2013).

(a) (b)

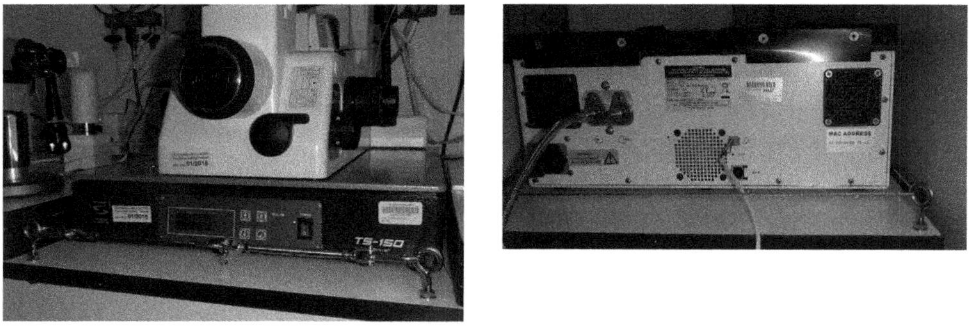

Fig. 12.1 Stabilizing equipment in a clinic where earthquake shudders are common. The strong wire protects against minor displacements. (a) ICSI station. TS150 JRS scientific instruments/The Table Stable Ltd. (b) Benchtop incubator. Planer PLC. (*Images kindly provided by Fertility Associates, Auckland, New Zealand.*)

is made up of two factors: severity and likelihood of frequency/recurrence. The earthquake case report above provides an excellent illustration of these principles:

- In terms of severity, an earthquake is clearly classified without hesitation as catastrophic or significantly catastrophic (4 or 5 on the scale of severity).
- The likelihood of frequency/recurrence is related to the geographical location of the clinic, and would be rated as 5 in Japan, and 1 in Belgium.

Evaluation of risk assessment/rating is not always easy, and a risk management team or working party has the task of enumerating risks and rating them by means of quantitative

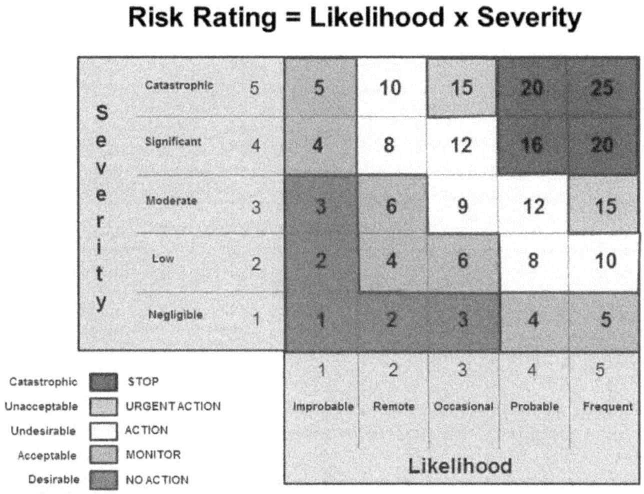

Fig. 12.2 Matrix used to assess risk factors on a scale of 1–5.

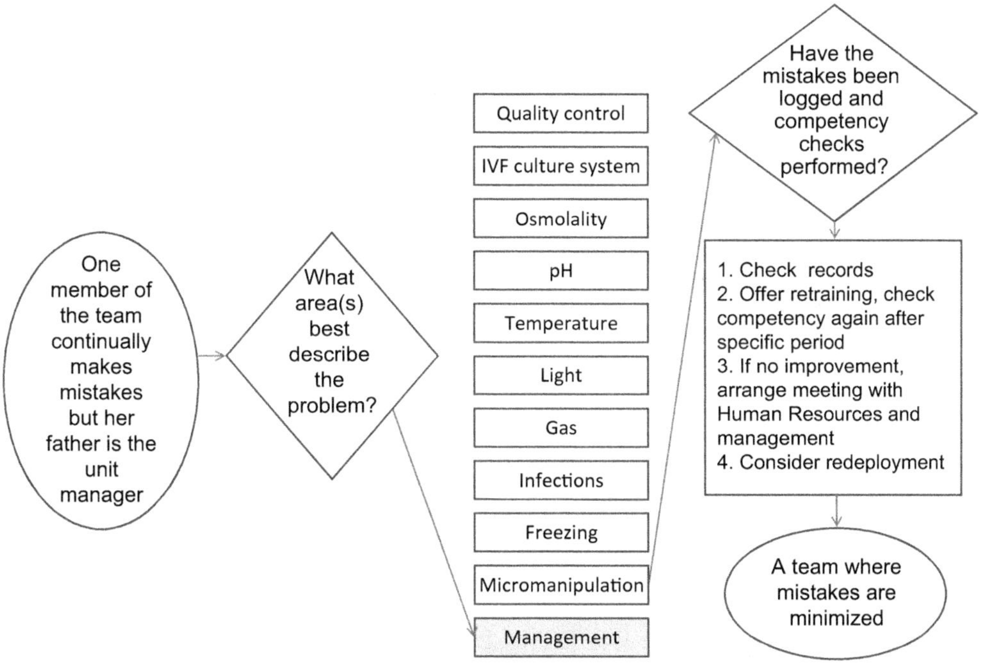

Fig. 12.3 Process map for troubleshooting management problems.

and repeatable metrics that will ensure a consistent method of evaluation. The major task of the group is to devise SOPs that define how events are to be handled. A risk matrix (Fig. 12.2) is a helpful tool, but should not be used in isolation: always remember that people are the drivers of the system.

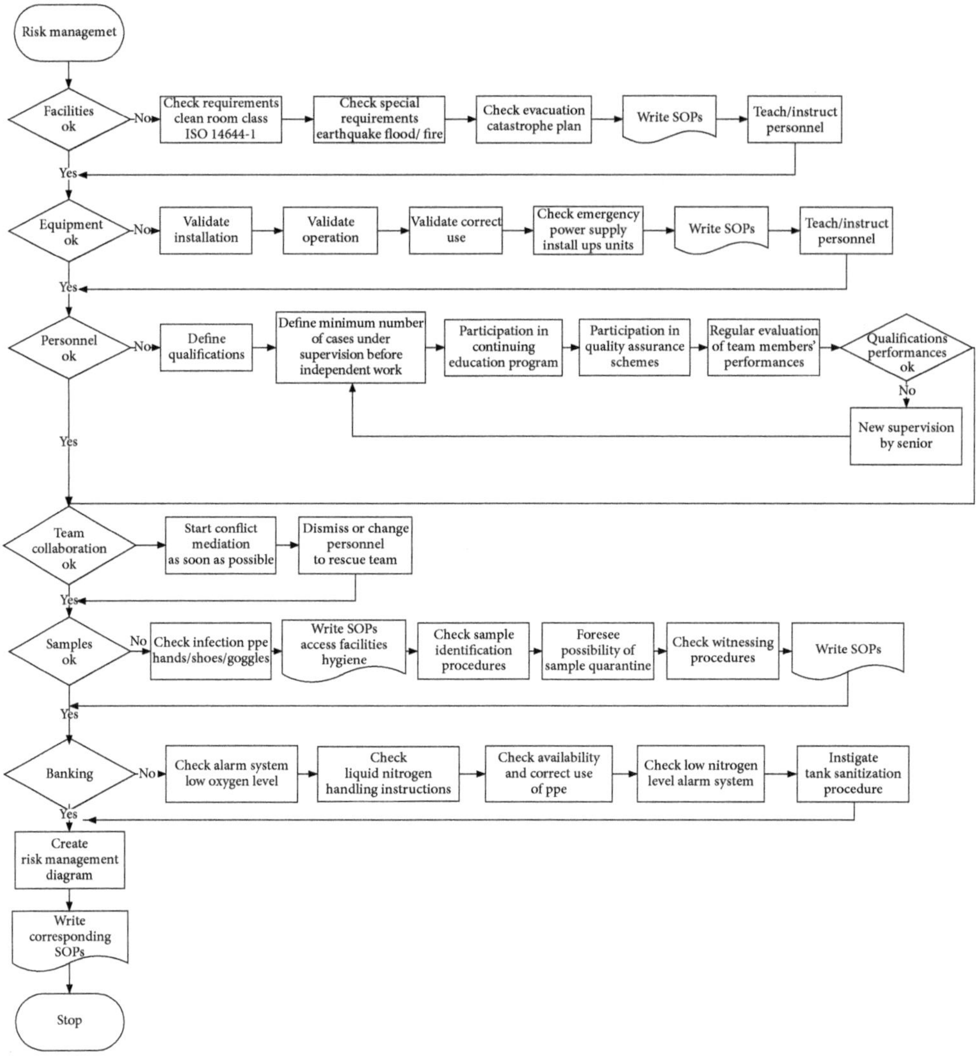

Fig. 12.4 Flowchart for troubleshooting management problems.

Conclusion

Once we truly know that ART is difficult – once we truly understand and accept it – then ART is no longer difficult, because once the principle is accepted, the fact that it is difficult no longer matters.

References

Choi Pak (2007). Multidisciplinarity, interdisciplinarity, and transdisciplinarity in health research, services, education and policy: 2. Promoters, barriers, and strategies of enhancement. *Clin Invest Med*, **30**(6), E224–32.

Glasl F (1999). *Confronting Conflict.* Hawthorn Press, Bristol.

Goldman KN, McCaffrey C, Ghosh M, Alder A, Chin A, Grifo JA (2013). No embryos left behind: emergent vitrification of 90 embryos during Hurricane Sandy. *ASRM poster.*

Goldman KN, McCaffrey C, Noyes N, Goldstein-Tufero A, Lee T-L, Grifo JA (2013). Blame it on the rain: salvaging IVF cycles threatened by Hurricane Sandy using emergency oocyte cryopreservation.

Human Fertilization and Embryology Authority (www.hfea.uk.gov) (2009). Code of Practice, 8th edn; www.hfea.gov.uk/docs/8th_Code_of_Practice.pdf.

Ishihara O, Yoshimura Y (2011). Damages at Japanese assisted reproductive technology clinics by the Great Eastern Japan Earthquake of 2011. *Fertil Steril*, **95**(8), 2568–70.

ISO 14971:2007 - Medical devices - Application of risk management to medical devices.

Karpman S (1968). Fairy tales and script drama analysis. *Transact Anal Bull*, 7(26), 39–43.

Practice Committees of the American Society for Reproductive Medicine and the Society for Assisted Reproductive Technology (2012). Recommendations for development of an emergency plan for *in vitro* fertilization programs: a committee opinion. *Fertil Steril*, **98**(1), e3–5.

Tomlinson MJ, Harbottle SJ, Woodward BJ, Lindsay KS; Association of Biomedical Andrologists (2012). Association of Biomedical Andrologists - laboratory andrology guidelines for good practice version 3 - 2012. *Hum Fertil (Camb)*, **15**, 156–73.

World Health Organization (2010). *WHO Laboratory Manual for the Examination and Processing of Human Semen*, 5th edn. World Health Organization Press.

Index

Lightning Source UK Ltd.
Milton Keynes UK
UKHW052209261118
332938UK00006B/427/P

9 781107 673175